Development of a Ship-Based CO$_2$ Transport Chain

I0131847

Development of a Ship-Based CO₂ Transport Chain

Vom Promotionsausschuss der

Technischen Universität Hamburg

zur Erlangung des akademischen Grades

Doktor-Ingenieur (Dr.-Ing.)

genehmigte Dissertation

von

Frithjof Engel

aus

Henstedt-Ulzburg

2019

Bibliografische Information der Deutschen Nationalbibliothek
Die Deutsche Nationalbibliothek verzeichnet diese Publikation in der
Deutschen Nationalbibliografie; detaillierte bibliographische Daten sind im Internet
über http://dnb.d-nb.de abrufbar.
1. Aufl. - Göttingen: Cuvillier, 2019
   Zugl.: (TU) Hamburg, Univ., Diss., 2019

1. Gutachter:	Prof. Dr.-Ing. Alfons Kather
2. Gutachter:	Prof. Dr.-Ing. Friedrich Wirz
Prüfungsausschussvorsitzender:	Prof. Dr.-Ing. Mathias Ernst
Tag der mündlichen Prüfung:	1. April 2019

© CUVILLIER VERLAG, Göttingen 2019
   Nonnenstieg 8, 37075 Göttingen
   Telefon: 0551-54724-0
   Telefax: 0551-54724-21
   www.cuvillier.de

Alle Rechte vorbehalten. Ohne ausdrückliche Genehmigung des Verlages ist
es nicht gestattet, das Buch oder Teile daraus auf fotomechanischem Weg
(Fotokopie, Mikrokopie) zu vervielfältigen.
1. Auflage, 2019
Gedruckt auf umweltfreundlichem, säurefreiem Papier aus nachhaltiger Forstwirtschaft.

   ISBN 978-3-7369-7000-7
   eISBN 978-3-7369-6000-8

Acknowledgments

Die vorliegende Arbeit entstand während meiner Zeit als wissenschaftlicher Mitarbeiter am Institut für Energietechnik der Technischen Universität Hamburg.

Ich bedanke mich bei Herrn Prof. Dr.-Ing. Alfons Kather für seine hervorragende fachliche Betreuung und persönliche Unterstützung. Seine Herangehensweise an die Lösung von Problemen und sein Einsatz für eine verständliche Darstellung komplexer Sachverhalte haben mich sehr geprägt.

Ebenso danke ich Herrn Prof. Dr.-Ing. Friedrich Wirz für die Übernahme des Zweitgutachtens sowie Herrn. Prof. Dr.-Ing. Mathias Ernst für die Übernahme des Prüfungsvorsitzes.

Den Kolleginnen und Kollegen, die ich während meiner Zeit am Institut kennenlernen durfte, danke ich für den freundschaftlichen Umgang untereinander und die vielen fachlichen Diskussionen. Besonderer Dank gilt meinem Büro- und Projektkollegen Lasse, den vier Jans und Philipp für die unvergessliche Zeit innerhalb und außerhalb der Institutsräumlichkeiten. Ich blicke auf viele gemeinsame Momente zurück und hoffe, dass wir die entstandenen Freundschaften noch bis zur Rente (und darüber hinaus) aufrechterhalten können.

Ich danke meinen Freuden, die mir in den vergangenen Jahren immer wieder gezeigt haben, dass es viele wichtige und interessante Dinge außerhalb der Ingenieurwissenschaften gibt.

Meiner Familie - insbesondere meiner Mutter - danke ich für die umfassende Unterstützung meines universitären und beruflichen Werdegangs.

Ganz besonders möchte ich mich bei Eva für die intensive gemeinsame Zeit und die emotionale Unterstützung bedanken. Ich freue mich auf unseren weiteren gemeinsamen Lebensweg.

Hamburg, im April 2019

Frithjof Engel

Table of Contents

List of Figures

III

List of Tables

Abbreviations and Symbols

Abbreviations

ASME	American Society of Mechanical Engineers
BOG	Boil-off gas
CAPEX	Capital expenditure
CCS	Carbon (dioxide) capture and storage
COP	Coefficient of performance
DNV	Det Norske Veritas (now: DNV GL)
GDP	Gross domestic product
GWP	Global warming potential
IPCC	Intergovernmental panel on climate change
LFL	Low flammability limit
LMTD	Logarithmic mean temperature difference
LNG	Liquefied natural gas
LPG	Liquefied petroleum gas
ODP	Ozone depletion potential
OPEX	Operating expenditure
PU	Polyurethane
TVL	Time weighted average threshold

Greek letters

λ	Thermal conductivity
Π	Pressure ratio
ρ	Density, without index: relative density
σ	Stress

Latin letters

A	Factor for design pressure calculation in the DNV rules, surface area for boil-off gas calculation
b	Tank width in the DNV rules
C	Refrigeration cycle compressor in the liquefaction process, characteristic tank dimension in the DNV rules
d	Tank diameter

e	Joint efficiency factor in the DNV rules
EX	CO_2 expander
h	Tank height in the DNV rules,
	(convective) heat transfer coefficient for boil-off gas calculation
H	Enthalpy
HE	CO_2/refrigerant heat exchanger
l	Length
$\dot{m}$	Mass flow rate
M	Mass
n	Number
p	Pressure
P	Mechanical or electrical power
PS	Refrigerant liquid-vapour phase separator
r	Tank radius
SW	Seawater-refrigerant heat exchanger
t	Wall thickness in tank calculation, time in transport chain model
T	Temperature
U	Overall heat transfer coefficient
Q	Heat quantity
$\dot{Q}$	Heat flow
z	Number of refrigeration stages

Indices

I, II, III	Refrigeration stage I, II, III (from upper to lower temperature)
0	Vapour pressure
A	Allowable dynamic membrane stress in the DNV rules
B	Tensile strength in the DNV rules
BOG	Boil-off gas
c	Compressor
cool	Cooling or refrigeration
cond	Condenser
evap	Evaporator
F	Yield stress in the DNV rules
gd, max	Internal liquid pressure due to dynamic loads in the DNV rules
i	Inner
in	Input
ins	Insulation
l	Liquid
min	Minimum
o	Outer
ofstor	Offshore intermediate storage

out	Output
p	Pipeline
s	Ship
spec	Specific
stor	Onshore intermediate storage
t	Tangential stress in the DNV rules
trans	Transport
v	Vapour
z	Number of compressor stages

1 Introduction

The impact of climate change on the future of humankind is a growing concern within society. The intergovernmental panel on climate change (IPCC) concludes that "continued emission of greenhouse gases will cause further warming and long-lasting changes in all components of the climate system, increasing the likelihood of severe, pervasive and irreversible impacts for people and ecosystems" [1]. Two of the 17 United Nations Sustainable Development Goals address the mitigation of climate change and the shift towards a low-carbon economy: Goal 7 demands universal access to affordable and clean energy. Goal 13 urges taking action to mitigate greenhouse gas emissions and combat climate change. Consequently, the provision of affordable, clean and reliable energy is one of the major global challenges for the next decades.

The rise in global CO_2 emissions can mainly be attributed to the continuous increase in electricity generation and the production of building materials such as cement and steel. Strategies to reduce CO_2 emissions from electricity generation and industry production can be divided into three different categories:

- a shift towards less CO_2-intensive primary energy sources such as natural gas, nuclear fuels and - in particular - renewable energy sources
- increased efficiency of electricity generation and industry production as well as increased end-use efficiency
- capture of the produced carbon dioxide and subsequent utilisation (carbon capture and utilisation – CCUS) or storage (CCS)

The International Energy Agency (IEA) projects that the majority of emission reduction can be obtained by a shift to renewable energy and increased end-use efficiency. However, it is assumed that in 2050, 17 % of the global electricity production will still be provided by fossil fuels [2]. Thus, carbon capture and storage is considered to be a key technology to achieve the two-degree goal agreed upon at the 21st United Nations climate conference in 2015. The deployment of CCS is also seen to be of particular importance for the industrial sector, since emissions from industrial production partly result from the chemistry of the process rather than from fuel combustion alone. Most computer models considered in the current IPCC report could not (likely) limit warming to below 2 °C if bioenergy, CCS and their combinations are limited. It is estimated that without CCS, the mitigation costs would increase by 138 % on average in the timeframe of 2015 to 2100 [1].

1.1 Motivation

A CCS infrastructure consists of CO_2 capture, transport and CO_2 storage. CCS research has historically been focused on CO_2 capture and CO_2 storage since transport is seen to be the least challenging part of the overall CCS chain [3]. Among other aspects, this can be attributed to the fact that there is significant experience with commercial CO_2 pipeline transport in the context of enhanced oil recovery [4]. However, knowledge gaps in CO_2 transport still exist and more research is necessary to ensure safe, reliable and economic CO_2 transport. Especially in the light of the public discussion about the future role of CCS, a thorough study of CO_2 transport is important.

From today's point of view, offshore CO_2 storage seems to be preferred by the general public over onshore CO_2 storage [5]. For large-scale CO_2 transport to an offshore geological storage location, either pipeline transport or ship transport are considered. The main benefits of pipeline transport are the available commercial experience and the low specific costs per quantity of CO_2 for small and medium transport distances. The main benefits of ship transport are its flexibility in terms of the connected CO_2 sources and sinks as well as in terms of the transported quantity. Moreover, the investment costs of ship transport are significantly lower than for pipeline transport. These benefits make ship transport an interesting alternative for offshore CO_2 transport, especially in the earlier stages of commercial CCS implementation.

Ship-based CO_2 transport is already used commercially on small scales in ammonia production and in the beverage industry [6]. The CO_2 is transported in liquefied state at low temperatures between -30 °C and -40 °C and pressures of up to 18 bar. For large-scale CO_2 transport, a lower temperature of -50 °C is usually recommended to reduce the pressure and thus, the investment costs of the CO_2 tanks. These temperatures are similar to commercial liquefied petroleum gas (LPG) transport, so that existing experience can be transferred.

While research on ship-based CO_2 transport in the context of CCS has already been conducted, these studies usually focus on one aspect, such as CO_2 liquefaction [7–12] or CO_2 injection [11, 13–20], rather than carrying out a thorough analysis of the overall transport chain. A concept design of an exemplary transport chain which includes the energy demand for liquefaction and injection as well as the necessary ship and intermediate storage capacities has not been conducted. Moreover, most studies assume a pure CO_2 feed stream rather than a realistic CO_2 stream composition which also includes various impurities from the fuel and combustion process. The CO_2 stream composition is mainly determined by the CO_2 capture process and its design and operating parameters [21–27]. Especially components with high boiling temperatures such as Hydrogen, Nitrogen and Oxygen can have a significant impact on the design and energy demand of the transport chain [6, 23, 28–30]. Thus, a study of the overall ship-based CO_2 transport chain and the impact of impurities on the transport chain components is necessary.

1.2 Aim and Scope

The aim of this work is the development of a ship-based CO_2 transport chain for CCS. The purpose is to provide an estimation for the energy demand and the dimensioning of the required components. Contrary to other works, a high-pressure CO_2 feed stream from a pipeline rather than a low-pressure feed stream from the capture plant is assumed. This configuration is based on a scenario where multiple CO_2 emitters are connected to a central pipeline for onshore transport and subsequent offshore transport to the geological storage location is desired. The results of this work are intended to serve as a basis for the comparison between ship-based and pipeline-based offshore transport. Particular focus is placed on the impact of typical impurities from different CO_2 capture technologies on the energy demand and design of the overall transport chain.

Models of the liquefaction process and the injection process are developed in Aspen Plus V8.6® to determine the specific energy demand per quantity of CO_2. An emphasis is put on the optimisation of the liquefaction process as it represents the main source of the energy demand for ship-based CO_2 transport. Sensitivity analyses are carried out to select the optimum operating and design parameters. The impact of certain measures of optimisation is studied. For the injection process, different heat sources such as seawater heat and engine waste heat are employed. Besides

3

liquefaction and injection, the specific energy demand for boil-off gas reliquefaction is determined. Alternative options for boil-off gas handling are considered. A model of the overall transport chain is developed to determine the necessary intermediate storage and ship capacities in dependency of the CO_2 mass flow rate as well as the overall energy demand for liquefaction, injection, and if desired, boil-off gas reliquefaction.

Three exemplary scenarios with different CO_2 quantities, impurity concentrations and CO_2 feed-in characteristics are studied – a 1 Mt/a scenario with a pure CO_2 stream, a 2 Mt/a scenario with an Oxyfuel CO_2 stream, and a 20 Mt/a scenario with different CO_2 emission sources. Sensitivity analyses are carried out to determine the influence of certain transport chain parameters such as the number of ships, ship capacity and the transport distance. The potential use of offshore intermediate storage for the decoupling of CO_2 transport and injection is evaluated.

2 Current State of Science and Technology

In the context of CCS, ship-based CO_2 transport is considered to be a potential alternative to pipeline transport. Studies suggest that ship transport is often more economical than pipeline transport for larger distances. The minimum transport distance for which ship transport becomes more economic than offshore pipeline transport (break-even point) depends on the transported quantity and the boundary conditions. Early literature on CO_2 ship transport found the break-even point to be between 500 km [31] and 1000 km [4] for a quantity of 10 Mt/a. Recent literature on the economics of CO_2 ship transport suggests that the break-even is usually lower, with values ranging between 200 km for a quantity of 2 Mt/a [32], 250 km for 2.5 Mt/a [33], and 300 km for 10 Mt/a [34]. Besides a potential cost advantage, the benefits of ship transport are its flexibility in terms of the transported quantity, CO_2 source and storage location. Moreover, a ship transport infrastructure can be implemented in a relatively short time period. For these reasons, ship-based CO_2 transport is considered to be a viable option in the early stages of commercial CCS deployment, even for shorter distances.

Figure 1 shows the individual components of the ship-based CO_2 transport chain considered in this work: A CO_2 stream from a pipeline is liquefied, stored at the harbor in an onshore intermediate storage, transported by ship and injected into a geological CO_2 storage located offshore. The CO_2 stream originates from one or more CO_2 sources such as power plants or large industrial CO_2 emitters, equipped with either Post-Combustion, Oxyfuel or Pre-Combustion CO_2 capture. During ship transport, a fraction of the CO_2 will evaporate, forming the so-called boil-off gas. Depending on the parameters of the transport chain, boil-off gas reliquefaction might be considered. At the geological CO_2 storage location, the CO_2 can either be directly injected from the ship (option A in Figure 1) or unloaded into offshore intermediate storage so that the ship can return to the harbour while injection is still continuing (option B).

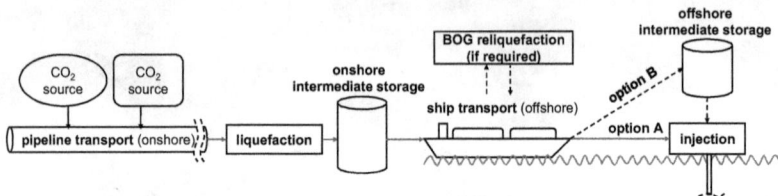

Figure 1:   The ship-based CO_2 transport chain considered in this work. The two methods for injection are direct injection from the ship (option A) or unloading into offshore intermediate storage (option B).

2.1 Liquefaction of CO_2 for Ship Transport

The two main process design principles are closed and open cycle CO_2 liquefaction. In the closed cycle process, an external refrigeration cycle separate from the CO_2 product stream is used for CO_2 liquefaction. In the open cycle process, a fraction of the CO_2 stream is recycled and acts as the working fluid of the refrigeration process.

2.1.1 Closed Cycle Process

The closed cycle process is usually based on a vapour-compression refrigeration cycle. In the vapour-compression cycle, cooling is provided by the evaporation of the refrigerant at a low pressure and temperature level. After evaporation the refrigerant is compressed to a higher pressure to allow the condensation of the refrigerant at a higher temperature level. After condensation, the refrigerant is expanded to evaporation pressure. In this adiabatic process, a certain fraction of the refrigerant is evaporated and in turn the temperature is lowered. This process is also termed flash evaporation. The continuous cycle of expansion, evaporation, subsequent compression and condensation allows heat absorption on a low temperature level and heat rejection on a high temperature level. The T-s diagram and the log p-h diagram of the ideal vapour-compression cycle are shown in Figure 2. The log p-h diagram is the most commonly used representation for refrigeration cycle thermodynamics since the major design parameters of a refrigeration cycle - the pressures and specific enthalpy differences in the evaporator and the condenser – can be read from the axes.

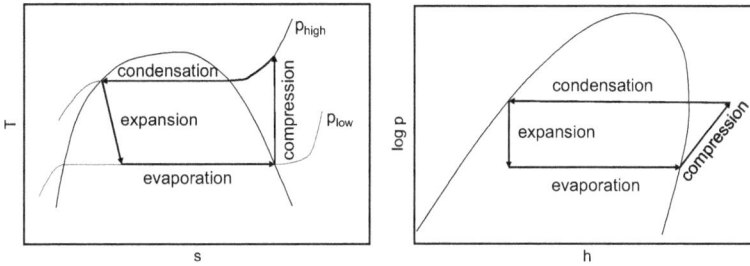

Figure 2: T-s diagram (left) and log p-h diagram (right) of the ideal vapour-compression refrigeration cycle

The efficiency of a refrigeration cycle is usually quantified by the coefficient of performance - COP:

$$COP = \frac{\dot{Q}_{cool}}{P_{in}} \tag{2.1}$$

The COP is the ratio of the cooling provided ($\dot{Q}_{cool}$) to the mechanical or electrical power input (P_{in}). In the case of the vapour-compression cycle, P_{in} is the power required for the refrigerant compressor(s). The value of COP primarily depends on the evaporation and condensation temperatures and pressures, the refrigerant type and the efficiency of the compressor(s).

A basic implementation of a closed cycle CO_2 liquefaction process with a vapour-compression refrigeration cycle is shown in Figure 3. The CO_2 is liquefied at a temperature of -50 °C. The liquefaction pressure depends on the impurity concentration and is 6.75 bar in the case of pure CO_2. The energy demand of this liquefaction process can be reduced by using multiple refrigeration cycles at different temperature levels. In this case, different refrigerants are often used for the upper and the lower temperature cycle (cascade configuration). A basic 2-stage closed cycle process with a cascade configuration is shown in Figure 4. While the total heat quantity transferred from the CO_2 stream to the refrigerant is the same as in the 1-stage closed cycle process, a certain fraction of the total heat can be transferred at a higher temperature in the upper temperature refrigeration cycle. The higher temperature leads to lower pressure in the upper cycle evaporator which reduces the overall compressor work in comparison to the 1-stage closed cycle process. Secondly, heat is rejected to ambience (e.g. seawater) only in the upper temperature cycle condenser, while the lower temperature cycle condenser transfers the heat to the upper temperature cycle. This means that the pressure ratio of the lower temperature cycle is lower than the pressure ratio of the compressor in a 1-stage closed cycle process. This effect contributes to the lower total energy demand of the 2-stage design.

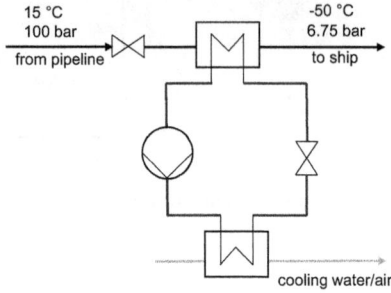

Figure 3:   1-stage closed cycle process for CO_2 liquefaction based on a vapour-compression refrigeration cycle

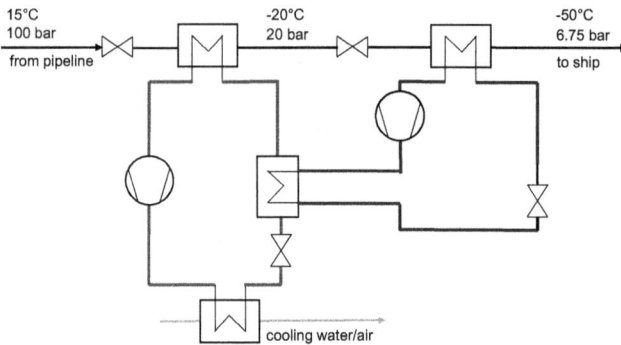

Figure 4:   The 2-stage closed cycle cascade process for CO_2 liquefaction. The temperatures and pressures apply to pure CO_2.

A side benefit of this multi-stage cascade design is that different refrigerants can be used for the upper and lower cycle which allows further optimisation in regard to investment costs, ease of handling and the necessary refrigerant volume. In industrial cooling, an NH_3-CO_2 cascade process is often used to avoid direct contact between ammonia and the refrigerated goods. Another advantage is that special safety precautions that come with the indoor use of ammonia can be avoided as the upper temperature ammonia cycle can be located outside the building while the lower temperature CO_2 cycle is situated inside. Moreover, the high density of CO_2 leads to lower space requirements inside the building.

When the use of different refrigerants is not required for operational reasons, the thermodynamically more advantageous multi-stage cycle without cascade configuration can be used. The 2-stage closed cycle process without cascade configuration is shown in Figure 5. Only one refrigerant is used and the cascade heat exchanger is replaced by a pressure vessel that acts as a liquid and gas phase separator. From the phase separator, the upper temperature cycle obtains gaseous phase and liquid phase refrigerant while the lower temperature cycle obtains the liquid phase one. Thus, the phase separator has the same effect as the cascade heat exchanger, i.e. condensation of the lower temperature cycle refrigerant and evaporation of the upper temperature cycle one. Unlike in a cascade process, a temperature difference between the upper and the lower cycle is not required. This means that the output pressure of the lower temperature cycle compressor is lower than in the cascade process, resulting in a lower energy demand. The lower temperature condensation pressure (and the upper temperature evaporation pressure) is usually set so that the compression ratio is the same for all stages [35]. The compression ratio for all stages can therefore be calculated by

$$\Pi_c = \left(\frac{p_{cond}}{p_{evap}}\right)^{1/z} \tag{2.2}$$

z is the number of stages. p_{cond} and p_{evap} denote the pressure in the refrigerant condenser and evaporator, respectively.

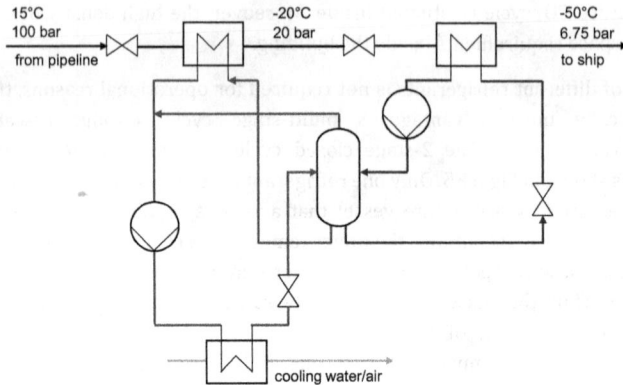

Figure 5: 2-stage closed cycle process for CO_2 liquefaction in which the cascade heat exchanger is replaced by a phase separator.

The closed cycle process for CO_2 liquefaction has already been studied by various authors. Abdulkarem et al. [8] used a 1-stage and a 2-stage cascade process for the liquefaction and compression of a pure CO_2 stream from 1 bar to 150 bar. Due to the different input and output conditions, their results cannot be directly transferred to the problem considered in this work. This also applies to the work of Øi et al. [7], who studied liquefaction and compression from 2 bar to 8 bar. In contrast, Decarré et al. [11] studied the liquefaction of CO_2 for ship transport which has previously been transported by pipeline – the same problem studied in this work. They determine a specific energy demand of 61 kJ/kg and 42 kJ/kg for a pipeline pressure of 100 bar and transport conditions of 7 bar, -50 °C and 15 bar, -30 °C, respectively.

In the context of CO_2 liquefaction, there is little information on refrigerant selection and - in the case of multi-stage processes - optimum values for intermediate pressures and temperatures. In industrial refrigeration, the following criteria are generally applied for refrigerant selection according to Hundy et al. [36]:

- A high latent heat of vaporization
- A high density of suction gas (i.e. at compressor inlet)
- Non-corrosive, non-toxic and non-flammable
- Critical temperature and triple point outside the working range
- Compatibility with component materials and lubricating oil
- Reasonable working pressures (not too high, or below atmospheric pressure)

10

- Low cost
- Ease of leak detection
- Environmentally friendly

No single refrigerant is optimum with respect to all these properties. Depending on the operating parameters and project requirements, usually one or more refrigerants are considered suitable. For CO_2 liquefaction, ammonia, ethane, propene and R134a have been used in the literature. Seo and Chang [37] calculated the COP of various multi-stage cascade processes for CO_2 liquefaction. Ethane was chosen for the lower temperature cycle and ammonia, propane and R134a were compared for the upper temperature cycle. Liquefaction temperatures lower than -30 °C, which are usually recommended for ship transport, were not investigated. Furthermore, an optimisation in regard to operating parameters such as the intermediate pressure was also not conducted.

2.1.2 Open Cycle Process

In contrast to the closed cycle process, in which an external refrigeration cycle is used for CO_2 liquefaction, the open cycle process is based on the concept of using the CO_2 stream itself for refrigeration. The open cycle process was first introduced by Aspelund and Jordal [9] for the liquefaction of a CO_2-rich stream from a CO_2 capture plant. In this original process, a CO_2 feed stream at 1 bar is compressed to 65 bar and sent to a distillation column for purification. Afterwards, the pressure is gradually reduced to 6.75 bar to reach the desired liquefaction temperature of -50 °C. At each intermediate pressure stage, a certain fraction of the CO_2 stream is recirculated, further expanded and used to refrigerate the CO_2 product stream. Lee and al. developed an improved version of the open cycle process [12, 38]. By optimising the internal energy recovery, a reduction in the energy demand of 10 % was obtained.

In contrast to the original process proposed by Aspelund and Jordal, a pure CO_2 feed stream is assumed in most other works on the open cycle process. Thus, purification - except for water removal in some cases – is not considered. A 1-stage open cycle process for the liquefaction of a pipeline CO_2 stream is shown in Figure 6. Starting at pipeline pressure, the CO_2 is expanded to a pressure of 57.4 bar. This pressure represents the boiling pressure at the CO_2 outlet temperature of the seawater heat exchanger (20 °C). After the recycling stream is added, the pressure is further reduced to the target pressure for ship transport (6.75 bar) to reach the desired liquefaction temperature of -50 °C. The principle of the open cycle process is similar to the closed cycle process with a vapour-compression refrigeration cycle, except that the CO_2 product stream itself is used as the refrigerant. In the open cycle process, the cooling duty is provided by the compression of the vaporous CO_2 fraction and the subsequent heat removal by seawater.

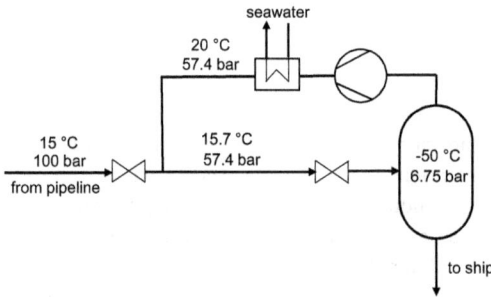

Figure 6: The 1-stage open cycle process for CO_2 liquefaction

Abdulkarem et al. [8] compared the open cycle process with the 1-stage and the 2-stage cascade closed cycle process. They found the energy demand of the closed cycle cascade processes (both 1-stage and 2-stage) to be lower than for the respective optimised open cycle processes, yet in the same order of magnitude. Seo et al. [39] carried out a technical and economic analysis of the open and closed cycle liquefaction processes. Feed stream conditions of 1 bar, 35 °C and the liquefaction conditions of 15 bar and -28 °C are assumed. With both lower OPEX and CAPEX costs, they found the closed cycle process to be more economic than the open cycle process. Similar results were obtained by Øi et al. [7] for a liquefaction pressure of 7 bar.

In the abovementioned works, a CO_2 feed stream at 1 bar is assumed. Consequently, the results in these works cannot be transferred to the scenario studied in this work, where a CO_2 stream from a pipeline is liquefied. This scenario has already been studied by Yoo et al. [34]: They use the 1-stage open cycle shown in Figure 6 as their base process and developed an improved 3-stage version. When using the improved open cycle process, they found the energy demand of combined pipeline and ship transport to be approximately 15 % higher than for pipeline transport only. A pure CO_2 feed stream is assumed, thus the impact of impurities on the feasibility of the process has not been studied.

2.2 CO2 Export Terminal

The CO_2 export terminal is the transfer point between the liquefaction plant and the CO_2 carrier ship. The export terminal consists of the onshore intermediate storage and the loading facilities at the berth where the ship is moored.

2.2.1 Onshore Intermediate Storage

The onshore intermediate storage functions as a buffer between the continuously operating CO_2 liquefaction process and the discretely operating loading process. The onshore intermediate storage consists of one or more thermally insulated pressure vessels. The temperature and pressure conditions are the same as for the CO_2 tanks on ships. The mechanical design of onshore intermediate storage tanks is carried out according to standard pressure vessel codes such as DIN EN 13445, AD 2000 Merkblätter or the ASME Boiler and Pressure Vessel Code. Suitable steels and insulation materials are the same as for CO_2 tanks on ships and are presented in section 2.3.

The capacity of the onshore intermediate storage depends on a number of parameters such as ship size, CO_2 mass flow rate from the liquefaction plant, transport distance, number of ships, ship speed, etc. The values for capacity given in literature are usually in the range of approximately 1 to 1.5 times the ship capacity. Lee et al. [38] found a capacity of 40,000 m^3 to be sufficient for a ship capacity of 30,000 m^3. Aspelund et al. [40], Barrio et al. [41], Engebø et al. [6] state that the onshore intermediate storage capacity should be 1.5 times the ship capacity. In another study by DNV GL, an onshore intermediate storage capacity of 20,000 t was found to be sufficient for a net ship capacity of 19,500 t [42]. However, in none of these studies is the calculation of the onshore intermediate storage capacity requirement documented, nor has a thorough analysis been carried out.

13

2.2.2 Loading

Flexible loading arms are widely used in the liquefied gas industry and can be used to connect onshore intermediate storage tanks with the CO_2 tanks aboard the ship. These loading arms are located directly on the jetty where the ship is moored and are designed to account for wave-induced motions of the ship during loading. Loading arm systems with pipe diameters between 8" and 24" are commercially available for the LNG industry [43]. A typical installation consists of three to five loading arms (two to four with the liquid and one with the gaseous product).

A potential connection between loading arms and onshore intermediate storage in the context of a liquefied CO_2 transport chain is shown in Figure 7. During loading of the ship, liquid CO_2 is transferred from onshore intermediate storage tanks to the ship via one or more loading arms. When the loading procedure starts, the CO_2 tanks aboard the ship mostly contain gaseous CO_2 at transport pressure except for a small trace of liquid CO_2 to avoid cavitation at the pump inlet. To maintain the transport pressure both in the CO_2 tanks aboard the ship and in the onshore intermediate storage tanks, the gaseous CO_2 is sent back from the ship to the onshore intermediate storage tanks via a separate loading arm. Due to the different densities, the mass flow rate of the liquid CO_2 is significantly higher than the mass flow rate of the recycled gaseous CO_2, e.g. by a factor of approximately 64 at -50 °C.

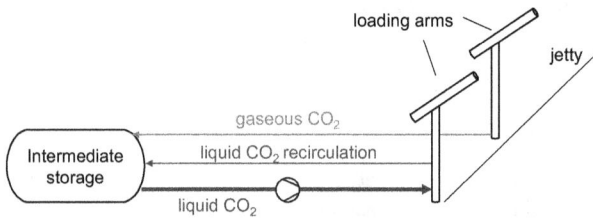

Figure 7:   Connection of loading arms and onshore intermediate storage. The gaseous CO_2 and liquid CO_2 lines are used during loading. The liquid CO_2 and liquid CO_2 recirculation lines are used during ship transport.

The loading mass flow rate primarily depends on the capacities of the onshore intermediate storage and the ship. In studies on CO_2 ship transport, loading mass flow rates of 500 t/h [15], 550 t/h [44] and 2083 t/h [42] have been selected. For Coral Carbonic, an existing CO_2 carrier with a capacity of 1375 t, a loading mass flow rate of 225 t/h has been used. For LNG transport, flow rates between 4,000 m^3/h and 15,000 m^3/h are commonly utilised [43], which correspond to CO_2 mass flow rates between 4600 t/h and 17250 t/h.

During ship transport, the loading arms are not in use. In that period, liquid CO_2 would evaporate inside the connection lines between the onshore intermediate storage and the loading arms, which would cause a significant pressure increase if no measurements were taken. A possible solution is to add a liquid CO_2 recirculation line (see Figure 7) which allows continuous flow inside the liquid CO_2 line [38]. This measurement enables evaporation to be avoided by using a recirculation mass flow rate which is rather low in comparison with the loading mass flow rate.

2.3 CO₂ Carriers

As of today, there are four CO_2 carriers commercially available, operated by two different companies. These CO_2 carriers are shown in Table 1. The Coral Carbonic was the first carrier specifically built for liquefied CO_2 transport. It is operated by Anthony Veder, a liquefied gas carrier chartering company. The ship is equipped with one tank, suitable for pressures of up to 18 bar and temperatures down to -40 °C. The other three carriers, Froya, Gerda and Embla, are of identical design and operated by Yara, a company that specialises primarily in ammonia and fertiliser production. The Yara CO_2 tanker fleet connects Yara production sites in Sluiskil (NL), Porsgrunn (NO), Fredericia (DK), Dormagen (DE) and Wilton (UK) to terminals in Hamburg (DE), Montoir (FR), Billingham (UK) and on the Thames (UK) [45]. These three ships replace an older fleet of two CO_2 carriers, each with 900 t cargo capacity operating at -30 °C and 20 bar [6].

Table 1:   Properties of commercially operated CO_2 carriers. The values marked by an asterisk (*) have been estimated based on the CO_2 density.

Name	Length in m	Width in m	Cargo capacity in t	Size of cargo tanks in m^3	Cargo conditions
Coral Carbonic	79.4	13.75	ca. 1375*	1250 m^3 L= 40 m, D_i= 6.4 m	up to 18 bar, down to -40 °C
Froya, Gerda, Embla	82.5	12.6	1800	ca. 1700*	up to 18 bar, down to -28 °C

While the pressure of the new Yara fleet is comparable to Coral Carbonic, the transport temperature is higher. However, it is mentioned in a publication by Yara representatives that for larger-scale CO_2 transport, it is most likely more economic to use a temperature near the triple point of CO_2 to decrease the transport pressure

15

[46]. With up to 18 bar, the maximum transport pressure of all four carriers is the same. This is above the boiling pressure of pure CO_2 for the considered transport temperatures: The boiling pressure of pure CO_2 is approximately 10 bar at -40 °C and approximately 15.3 bar at -28 °C, which means that both the Yara fleet and Coral Carbonic have a certain margin to the boiling pressure of pure CO_2. Since the boiling pressure increases if impurities are present, the Coral Carbonic carrier has a larger tolerance than the Yara fleet regarding impurities in the CO_2 cargo.

In the context of CCS, many authors suggest carriers in the range of 10,000 m³ to 50,000 m³ capacity for CO_2 ship transport [40, 44, 47–49]. Several concept designs for these CO_2 carriers already exist in literature and most of them are based on the conventional liquefied gas carrier design. Typically, 6 tanks are placed inside the ship's hull, two in parallel and 3 in a row with respect to the longitudinal axis of the ship. An overview of these concept designs is shown in Table 2. With total capacities between 23,000 t and 46,000 t, these carriers are considerably larger than the existing ones shown in Table 1. In contrast to the existing carriers, a lower transport temperature of -50 °C to -52 °C, i.e. just above the triple point temperature of CO_2, is chosen to attain low transport pressures between 6.5 bar and 7 bar.

Table 2: **Properties of concept designs for CO₂ carriers within the context of CCS. The values marked by an asterisk (*) have been estimated. The values marked by an '?' are unknown.**

Name	Length in m	Width in m	Cargo capacity in t	Configuration and size of cargo tanks in m³	Cargo conditions
Yoo et al. [34]	?	?	26,450*	2x3 tanks a 3.833 m³* with L= 35 m, D$_l$= 11 m	6.7 bar, -50 °C
Vesterdal et al. [50]	228	31	46,092*	2x3 tanks a 6680 m³ with L= 50 m, D$_l$= 13.5 m	6 bar -52 °C
Vermeulen [51]	210	11	34,500*	2x3 tanks a 5000 m³	7 bar -50 °C
Aspelund et al. [40]	?	?	23,000*	2x3 tanks a 3333 m³	6.5 bar -52 °C

Apart from the concept designs shown in Table 2, other preliminary designs exist. Omata et al. [15] studied a ship with a capacity of 2950 t and bilobe-tanks instead of cylindrical ones. It uses a temperature of -10 °C and a pressure of 30 bar. Yoo et al. [34] developed an alternative design for very large CO_2 carriers with a capacity of 110,000 m³. They suggest using a large number of small tanks placed next to each other in an upright position. Since both concepts are significantly different to the existing commercial CO_2 carriers and have not thus far been adopted by other authors, they are not further analysed in this work.

The construction and equipment of CO_2 carriers is regulated by the *International Code for the Construction and Equipment of Ships Carrying Liquefied Gases in Bulk* (IGC-Code) [52] issued by the International Maritime Organisation (IMO). This code contains, among other aspects, requirements for structural analysis and fabrication as well as safety regulations for environmental conservation and fire protection. The mechanical design of CO_2 carriers is carried out according to a specific set of classification rules, e.g. the "rules for the classification of ships" by DNV GL [53]. In the DNV GL rule framework, a separate chapter on liquefied gas carriers exists in which CO_2 carriers are explicitly mentioned as an example. A distinction is made between fully refrigerated, semi-pressurised and pressurised liquefied gas carriers. In fully refrigerated liquefied gas carriers, the cargo is transported at a low temperature and atmospheric pressure. This is suitable for cargo which can be liquefied at atmospheric pressure such as LPG or ammonia. In fully pressurised ships, the cargo is liquefied by compression instead of refrigeration. The semi-pressurised ship employs both compression and refrigeration for liquefied gas transport. Consequently, all existing CO_2 carriers in Table 1 are categorised as semi-pressurised ships.

The DNV classification for liquefied gas carriers categorises storage tanks into independent type A, type B and type C tanks [53]. The term 'independent' refers to the fact that these tanks are independent of the ship's hull structure, unlike integral or membrane tanks, for example, which use the inner hull structure for load bearing. CO_2 cargo tanks are categorised as so-called independent type C tanks, which means that they are designed as structurally independent cylindrical pressure vessels. Type A tanks, on the other hand, consist of flat surfaces and have a low operating pressure (< 0.7 barg). Consequently, they are used for fully refrigerated ships. Type B tanks have the same operating pressure limit but are subject to more detailed structural analysis, which includes, among other aspects, crack propagation and fatigue failure calculations. Type C tanks are constructed using the general DNV rules for pressure vessels [54] in combination with the rules for liquefied gas carriers [53]. Type C cargo tanks are used for semi-pressurised and pressurised ships. In contrast to common pressure vessel calculations, in which the internal pressure is equal to the vapour pressure of the liquefied gas, the internal pressure of type C cargo tanks results from the vapour pressure and additional dynamic loads from sea swell. In addition to the pressure vessel calculations presented in the following, a so-called buckling evaluation is necessary. The buckling evaluation ensures that the cargo tank is sufficiently designed against plastic deformation (or 'buckling') due to external forces. The calculations for the buckling evaluation are documented in the DNV rules [53, 55] and are omitted here for brevity.

17

Type C tanks can be considered as thin-walled since the diameter is significantly greater than the wall thickness. Thus, relatively simple formulas for the calculation of the cylindrical shell and the dished ends can be used. The greatest wall thickness is necessary for the cylindrical shell. It can be calculated from the following equation [55]:

$$t = \frac{p_{design} \cdot r_i}{10\sigma_t \cdot e - 0.5 \cdot p_{design}} \tag{2.3}$$

r_i is the inner radius of the tank. The factor e represents the joint efficiency factor and is equal to 1 for type C tanks. σ_t is the membrane stress and is calculated from the type and properties of the tank material. For the relevant low temperature steels, σ_t is the minimum of the following two quotients:

$$\sigma_t = \min(\tfrac{\sigma_B}{3}, \tfrac{\sigma_F}{2}) \tag{2.4}$$

σ_B is the tensile strength and σ_F the yield stress of the tank material. The design pressure p_{design} is the sum of the internal vapour pressure p_0 and the maximum combined internal liquid pressure $(p_{gd})_{max}$ due to dynamic loads such as gravity and acceleration.

$$p_{design} = p_0 + (p_{gd})_{max} \tag{2.5}$$

The internal vapour pressure p_0 is defined by

$$p_0 = 2 + A \cdot C \cdot \rho^{1.5} \tag{2.6}$$

The factor A considers the membrane stress σ_t and the allowable dynamic membrane stress $\Delta\sigma_A$, which is 55 N/mm^2 for any type of steel:

$$A = 0.0185 \left(\frac{\sigma_t}{\Delta\sigma_A}\right)^2 \tag{2.7}$$

The factor C represents the characteristic tank dimension, which is defined as the greatest of the tank height h, 0.75 of the tank width b or 0.45 of the tank length l:

$$C = \max(h, 0.75b, 0.45l) \tag{2.8}$$

For cylindrical tanks, h and b are equal to the diameter of the tank. ρ is the relative density of the cargo, which is the absolute cargo density at transport temperature divided by the density of fresh water at 4 °C:

$$\rho = \frac{\rho_{\text{cargo}}(T_{trans}, p_{trans})}{1000 \text{ kg/m}^3} \tag{2.9}$$

$(p_{gd})_{\text{max}}$ is calculated from the liquid height in the tank and the dimensionless acceleration resulting from gravity and dynamic acceleration effects. The formulas are omitted here for brevity. For the acceleration components, formulas are given for rough sea conditions in the North Atlantic occurring with a probability of less than 10^{-8}, which can also be used for other parts of the world to avoid individual calculations. The other relevant parameters for $(p_{gd})_{\text{max}}$ are calculated from the geometry of the tank. The result for $(p_{gd})_{\text{max}}$ is usually much smaller than the internal vapour pressure p_0. For example, a calculation of $(p_{gd})_{\text{max}}$ for an LPG cargo tank from literature [56] yields the value 1.44.

Preliminary calculations for the design of liquefied CO_2 cargo tanks have been conducted. The main limitations for liquefied CO_2 tank construction are dictated by the properties of the material. Special low temperature steels are required to avoid embrittlement at low temperatures. Usually, manganese or nickel-alloyed steels are used to reach the required ductility at low temperatures. Common steels for low temperature applications and their properties are shown in Table 3. These steels represent recommendations by AD 2000 Merkblatt W 10 [57] for low temperature pressure vessel steels. 13MnNi6-3 has a minimum temperature of -60 °C and 26CrMo4-2 has a minimum temperature -65 °C. Both steels are therefore suitable for CO_2 transport at a temperature of -50 °C. The X8Ni9 steel is a highly nickel-alloyed steel for cryogenic applications such as liquefied natural gas (LNG) transport at -160 °C. It is most likely not economic to use it for CO_2 transport, but it is included in Table 3 to illustrate that steel types with even higher yield stress and tensile strength properties are available. For the considered steels, the allowed membrane stress σ_t calculated according to equation (2.4) was found to be between 168 N/mm² and 217 N/mm².

Table 3:   Low temperature steels suitable for liquefied CO_2 tank construction [57]

Steel type	Yield stress σ_F in N/mm²	Tensile strength σ_B in N/mm²	Allowable membrane stress σ_t in N/mm²	Minimum temperature in °C	Maximum thickness in mm
13MnNi6-3	335	515	168	-60	50
26CrMo4-2	440	590	197	-65	40
X8Ni9	500	650	217	-200	50

Within the scope of the AD 2000 standard, the wall thickness is usually limited to 50 mm or even 40 mm for pressure vessels in low temperature regime. Figure 8 shows the wall thickness of the cylinder shell with an inner radius r_i of 4710 mm calculated by eq. (2.3) in dependency of the design pressure for three different allowable membrane stresses. These allowable membrane stresses are loosely based on the values from Table 3. The tank length represents the characteristic length C in eq. (2.8). The results in Figure 8 show that the necessary wall thickness of the cylinder shell increases with increasing design pressure and decreases with increased allowable membrane stress. For an assumed maximum wall thickness of 50 mm, the maximum design pressure is in the range of approximately 18.5 bar to 26.5 bar.

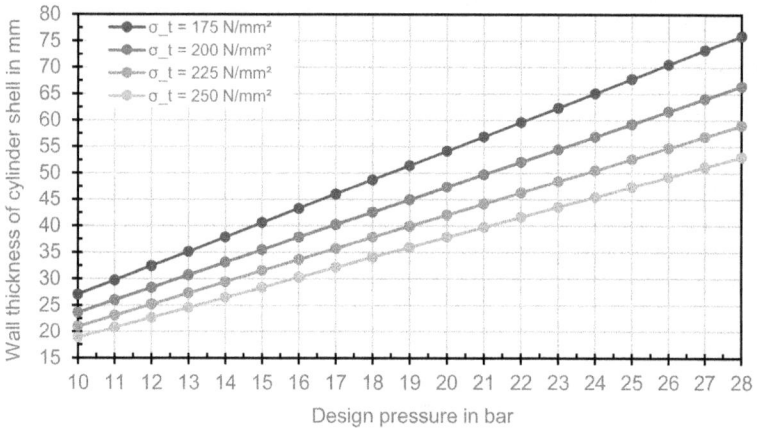

Figure 8: Wall thickness of the cylinder shell with an inner radius r_i of 4710 mm according to eq. (2.3) in dependency of the design pressure p_{design} for different allowable membrane stresses σ_t

For a given wall thickness, a given design pressure and a given allowable membrane stress, the maximum inner diameter of the tank can be determined according to eq. (2.3) In Figure 9, the maximum inner tank diameter d_i is shown in dependency of the design pressure for a wall thickness of 50 mm. The results show that the maximum inner tank diameter decreases with increased design pressure. As a rule of thumb, it can be said that within the considered range, the diameter is approximately halved when the design pressure is doubled. For example, the maximum inner diameter for a design pressure of 14 bar and an allowable membrane stress of 175 N/mm^2 is 6.2 m while it is 3.1 m for a design pressure of 28 bar.

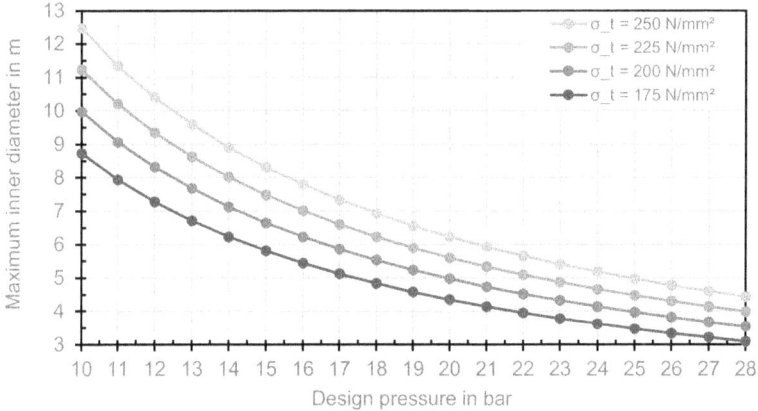

Figure 9: **Maximum inner diameter of tank according to eq. (2.3) in dependence of the design pressure for different allowable membrane stresses σ_t and a wall thickness of 50 mm**

For a given tank length, a decrease in the maximum inner diameter implies a decrease in the maximum tank volume. Figure 10 shows the volume of a cylinder with a length of 50 m in dependency of the design pressure. For the inner diameter, the values from Figure 9 were used. The cylinder volume is a conservative estimate of a real cargo tank volume which would also include the volume of the dished ends. The values in Figure 10 show a considerable decrease in the cylinder volume with decreasing inner diameter, which is a result of the quadratic relationship of the inner diameter and the cylinder volume. Thus, a doubling of the design pressure leads to a reduction in the volume by a factor of four. This directly corresponds to the number of cargo tanks that are required to transport a certain volume. Therefore, the investment costs for liquefied CO_2 ship transport significantly increase with increased transport pressure. Consequently, a low transport pressure is usually recommended in literature (q. v. section 3.1).

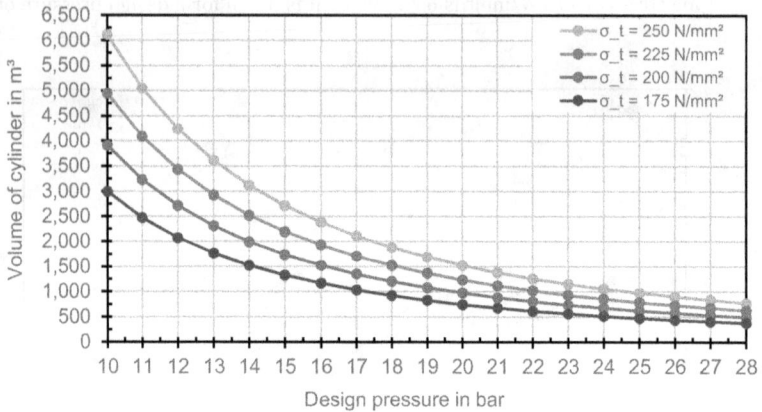

Figure 10: Volume of a cylinder with a length of 50 m and an inner diameter according to Figure 9

The maximum inner diameters shown in Figure 9 and the tank volumes shown in Figure 10 are in good agreement with the commercially operated CO_2 carriers shown in Table 1: The design pressures of these CO_2 carriers are estimated to be between 19 bar and 20 bar (the transport pressure of 18 bar plus $(p_{gd})_{max}$, which is typically between 1 bar and 2 bar). As the transport temperatures are higher than -50 °C, higher allowable membrane stresses than the ones shown in Table 3 are attainable. This means that the values shown in Figure 9 and Figure 10 are also realistic for a tank length of 40 m, i.e. for the Coral Carbonic. Likewise, the higher tank volume of the Yara fleet is realistic considering the higher transport temperature of -28 °C.

For the concept designs shown in Table 2, inner tank diameters between 11 m and 13.5 m have been selected. The design pressure is estimated to be in the range of 7 bar to 9 bar. For these design pressures, the maximum inner tank diameter calculated according to eq. (2.3) is in the range of 9.7 m and 17.8 m, depending on the allowable membrane stress. Therefore, the inner diameters of the concept designs are in alignment with the DNV rules. Results for design pressures lower than 10 bar are not shown in Figure 9 and Figure 10 to improve the readability of the diagram for higher design pressures. The cylinder volume for an inner diameter between 9.7 m and 17.8 m is in the range of approximately 3700 m³ to 12400 m³ for a tank length of 50 m. These values are in alignment with the 6680 m³ tank volume in the concept design by Vesterdal et al. [50], for which the same tank length is used.

22

Equivalently, these values are in good accordance with the 3833 m³ in the design by Yoo et al. [34], scaled down to 70 % to account for a tank length of 35 m. The inner diameter and the tank length for the concept designs by Vermeulen [51] and Aspelund et al. [40] are unknown, but similar tank volumes are employed. Therefore, it can be concluded that calculation of the tank geometry according to the DNV rules is in good agreement with the tank geometries of the concept designs shown in Table 2.

CO_2 tanks are insulated to minimise the heat ingress from ambience. Equivalent to other liquefied gas tanks, either Polyurethane (PU) foam or evacuated perlite is usually considered. With around 0.02 W/mK, PU foam has a lower thermal conductivity than evacuated perlite with values around 0.035 W/mK. Therefore, the thickness of the insulation layer must be higher in the case of evacuated perlite if the same thermal resistance is desired. In literature on CO_2 tanks, an insulation layer thickness of 0.15 m has been chosen for PU foam [50] and a thickness of 0.30 m for evacuated perlite [58]. Figure 11 shows the overall heat transfer coefficient for the cylinder section of a tank in dependency of the insulation layer thickness.

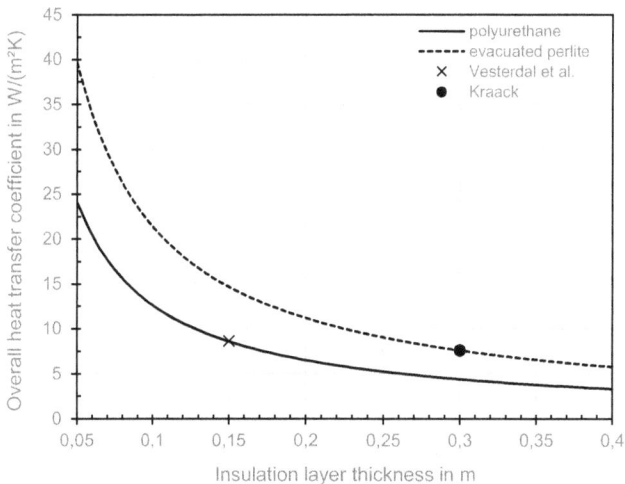

Figure 11: Overall heat transfer coefficient for the cylinder section of a CO2 tank in dependency of the insulation layer thickness. For the steel layer, a thickness of 50 mm and a thermal conductivity of 25 W/mK are assumed. The convective heat transfer coefficients are 10 W/mK for the air side and 100 W/mK for the liquid CO2 side. Literature values from Vesterdal et. al [50] and Kraak [58] are considered.

23

For the steel layer, a thickness of 50 mm and a thermal conductivity of 25 W/mK are assumed. The convective heat transfer coefficient for the air side is 10 W/mK and 100 W/mK for the liquid CO_2 side [50]. Figure 11 illustrates that the marginal utility of an additional centimetre of insulation layer in regard to the overall heat transfer coefficient becomes increasingly lower with increased insulation layer thickness. The comparison of the two available literature sources shows that similar overall heat transfer coefficients are obtained, even though different materials and different insulation layer thicknesses have been used: With 0.15 m of PU foam as used by Vesterdal et al. [50], the overall heat transfer coefficient is 8.6 $W/(m^2K)$, while it is 7.6 $W/(m^2K)$ for an insulation layer of 0.30 m evacuated perlite such as that used by Kraak [58].

2.4 Boil-off Gas

Although CO_2 tanks are insulated, heat ingress from ambience cannot be completely avoided. As a result, a small fraction of the CO_2 is evaporated, forming the so-called boil-off gas (BOG). If the boil-off gas is retained inside the tanks, the pressure is increased, which may not be tolerable if a significant amount of BOG is formed. Consequently, if a constant pressure in the tank is desired, the boil-off gas must be vented and either released into the atmosphere or reliquefied and sent back to the tank. The usual order of magnitude for boil-off gas rates of CO_2 tanks given in literature is 0.1 %/d (mass-%) [59]. A value of 0.15 %/d is assumed by Yoo et al. [34]. Other literature sources find it to be higher by a factor of 10 [4], although it is not clear whether the CO_2 emissions of the ship's propulsion are included. Other sources find it to be as low as 0.01 %/d [50]. A student's thesis conducted on this subject suggests a value of 0.1 %/d for large scale CO_2 tanks with a capacity of 5000 m^3[60].

Boil-off gas handling has mainly been studied in the context of liquid fuel transport such as LPG and LNG [61, 62]. The main difference between liquid fuel boil-off and CO_2 boil-off gas is that vented LNG and LPG can be used for ship propulsion while CO_2 cannot. Even if the engine is not capable of utilising LNG or LPG, the boil-off gas is still a valuable good so boil-off gas reliquefaction is economic in most cases. LNG boil-off gas reliquefaction is an established technology and was patented in 1974 [63]. In contrast, CO_2 does not have an inherent value, so reliquefaction would only be feasible for high CO_2 prices or if it is required by legislation. Despite this fact, a few studies on boil-off gas in the context of CO_2 have already been carried out. A reliquefaction process was developed by Chu et el. [59], who calculated a COP of 1.6

to 2.1, depending on the liquefaction temperature. In another study [64], the impact of volatile components such as oxygen and nitrogen is analysed. Due to the boundary conditions selected in this study, the results cannot be directly compared to the present work.

In the liquefied CO_2 ship transport chain, boil-off gas is formed in the onshore intermediate storage tanks and in the CO_2 tanks aboard the ship. In Figure 12, the different strategies for boil-off gas reliquefaction are illustrated. Boil-off from the onshore intermediate storage tanks can be sent back to the liquefaction plant so that a separate reliquefaction process is not necessary (Figure 12a). Boil-off from the CO_2 tanks aboard the ship can either be vented or reliquefied using a dedicated process. As shown in Figure 12, this can potentially be a vapour-compression cycle (Figure 12 b). Alternatively, the Linde-Hampson process can be used (Figure 12 c). In this process, the boil-off gas is compressed, cooled by seawater and the incoming boil-off gas stream, expanded, and sent back to the tank.

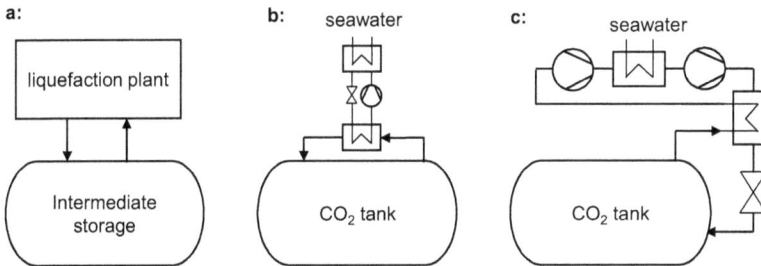

Figure 12: Reliquefaction strategies for boil-off gas from onshore intermediate storage tanks (a) and from CO_2 tanks aboard the ship using a vapour-compression cycle (b) and the Linde-Hampson process (c)

2.5 Offshore Injection

The main purpose of the injection process is to increase the CO_2 pressure from tank pressure to injection pressure, or more precisely, wellhead pressure. In Figure 13, the principle structure and components of a CO_2 injection process are shown. The process consists of a CO_2 pump, a seawater heat exchanger and potentially, an additional heat exchanger if seawater heat is not sufficient as the only heat source. As the CO_2 is in liquid state, a pump can be used instead of a compressor. After increasing the pressure of the CO_2, the temperature must be increased to avoid hydrate formation. There is no consensus in literature on the minimum CO_2 temperature at the wellhead: In two literature sources, a temperature of 0 °C is assumed to be sufficient [11, 17]. In another source, a wellhead temperature of -7 °C is used [34]. In both cases, ice formation around the injection pipeline or hose must be taken into account. Contrary to these literature sources, other authors assume that the minimum temperature to avoid hydrate formation is 10 °C [18] or 15 °C [13]. The work of Omata et al [15] contains a detailed analysis of the injection process carried out by Chiyoda, a Japanese company specialized in offshore oil and LNG industry. They assume a temperature of 5 °C at the wellhead.

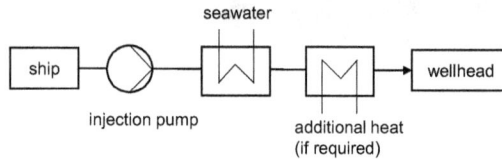

Figure 13: **Principle structure and components of a CO_2 injection process**

Depending on the literature source, different options for CO_2 heating are considered. Decaree et al. [11] assume seawater heat to be sufficient. Instead of using heat exchangers, they propose a long injection pipeline in which the CO_2 is heated by the surrounding seawater. They calculate the minimum pipeline lengths to be between 2 km and 7 km depending on the specific scenario. Other literature sources suggest that an external heat supply is necessary. Krogh et al. [18] and Aspelund et al. [13] consider external heat supply to be necessary for the North Sea region as the assumed seawater temperature of 5 °C is not high enough.

Only one literature source could be found in which the electrical and thermal energy demand of the CO_2 injection process is calculated [18]. In this article, a simulation of a 2-stage injection process was carried out. In the first step, the CO_2 input stream at -53 °C and 8 bar is compressed and heated to 26 bar and -30 ° C. In the second step, the CO_2 is further compressed and heated to 124 bar and 0 °C for injection. The electrical energy demand for this process was determined to be approximately 4 kWh/t CO_2 and with a thermal energy demand of 25 kWh/t CO_2.

Three options exist for the offshore installation of the injection process components such as the injection pumps and the seawater heat exchanger. The advantages and disadvantages of these options are shown in Table 4. All three options have been commercially used in the oil and gas industry. A fixed platform can be used on which all components are installed. The ship is moored to the platform for unloading. The platform itself is connected to the well by a pipeline. An advantage of this setup is that only one injection pump and one seawater heat exchanger is required, regardless of the number of ships. A disadvantage is the high number of components that have to be installed offshore. Consequently, installation costs of this setup are quite high. It is also the most vulnerable system among the considered options with respect to rough sea conditions and potential ship collisions.

Table 4:   Potential systems for offshore CO_2 unloading and injection

	Platform	Ship	Barge
Advantage	only one injection pump and seawater heat exchanger	very little permanent offshore installation necessary	decoupling of unloading and injection mass flow rates, only one injection pump and seawater heat exchanger
Disadvantage	high vulnerability to rough sea conditions and collision	injection pump and seawater heat exchanger required on each ship	most likely only economic when a large number of ships is required

The second option is to inject the CO_2 directly from the ship so that no platform is required. In this case, the injection pump and the seawater heat exchanger are installed aboard the ship. For injection, the flowline (a pipeline or a hose) is connected to the ship and the ship is kept in position by mooring to the buoy to which the flowline is connected. For the connection of the ship and the flowline, different technologies such as buoy turret loading (BTL), Catenary Anchor Leg Mooring (CALM) or socket buoy systems are commercially available. In these systems, the flow line is kept at the sea surface even if not in use. This is a disadvantage if rough sea conditions are frequently expected. In this case, a pickup buoy technology such as the submerged turret loading (STL™) system can be used as an alternative. In these systems, a flexible pipe is kept on the seabed while not in use. For unloading and injection, the hose is raised via a separate pickup wire rope and connected to the ship. The ship is not moored to any permanent structure and is kept in position by a dynamic positioning system (DPS). The advantage of the direct injection methods is that very little permanent offshore installation is necessary. The disadvantage is that an injection pump and heat exchangers are required on every ship.

The third option is to permanently moor a ship (barge) at the injection site and use it as an offshore intermediate storage. This is sensible when the injection mass flow rate is comparably low so that injection without an offshore intermediate storage would require a lot of time. With offshore intermediate storage, unloading of the transport ship and the actual injection can be decoupled so that the time required for unloading is not determined by the injection mass flow rate and can therefore be minimised. As the injection pump and heat exchangers can be installed on the barge, the main disadvantage of not using a platform - i.e. requiring injection components on each ship - would be removed. Similar concepts are already used in oil and gas industry and are known as floating storage and offloading (FSO) solutions. A potential fourth option would be to install offshore intermediate storage on a platform, but this concept is most probably not economical in comparison to a barge. The main disadvantage of using a barge is that an additional ship with the same capacity as the transport ships is required. Thus, a barge is most likely only economic when the total number of ships is large so that the savings in unloading time and injection equipment outweigh the additional cost of the barge.

3 Model Development

The individual components of the ship transport chain that require energy input – liquefaction, boil-off gas reliquefaction and injection - have been modelled and optimised in regard to the specific energy demand per quantity of CO_2. Particular focus was placed on the optimisation of the CO_2 liquefaction process since it requires the highest energy input among all components of the ship transport chain (apart from CO_2 capture). In the next step, a model of the entire transport chain has been developed to allow the dimensioning of the individual components in dependency of the CO_2 amount transported per year.

3.1 CO_2 Stream Composition and Transport Conditions

Ship transport of CO_2 is carried out in liquefied state to attain a low transport pressure. The lower pressure and temperature limit for the liquid state of pure CO_2 are determined by the triple point at -56.6 °C and 5.18 bar where solid CO_2 ("dry ice") is produced. The upper temperature and pressure limit is determined by the critical point at 31 °C and 73.7 bar. Between these temperature limits, pure CO_2 exists in liquid state when the pressure of the CO_2 is higher than the boiling pressure at the respective temperature. When the CO_2 stream[1] contains impurities, the pressure must be higher than the so-called bubble pressure. The bubble pressure of a multi-component mixture is defined as the lowest pressure at which a gas bubble is formed at the given temperature. It depends on the boiling pressures of the

---

[1] In this work, a *CO₂ stream* is defined as a (potentially multi-component) gas, liquid or two-phase mass flow which predominantly consists of CO_2 but might also contain other components which are usually termed impurities. This simplification was done as this work deals with the transport of CO_2 streams rather than the CO_2 capture process, for which differentiation between CO_2-rich and CO_2-lean streams would be necessary.

individual components of the CO_2 stream. Thus, the bubble pressure of a single-component substance is equal to its boiling pressure.

In Figure 14 the bubble pressures are shown in dependency of the bubble temperature for typical CO_2 streams from Post-Combustion, Oxyfuel and Pre-Combustion CO_2 capture plants. Additionally, the boiling pressure and density of pure CO_2 are shown.

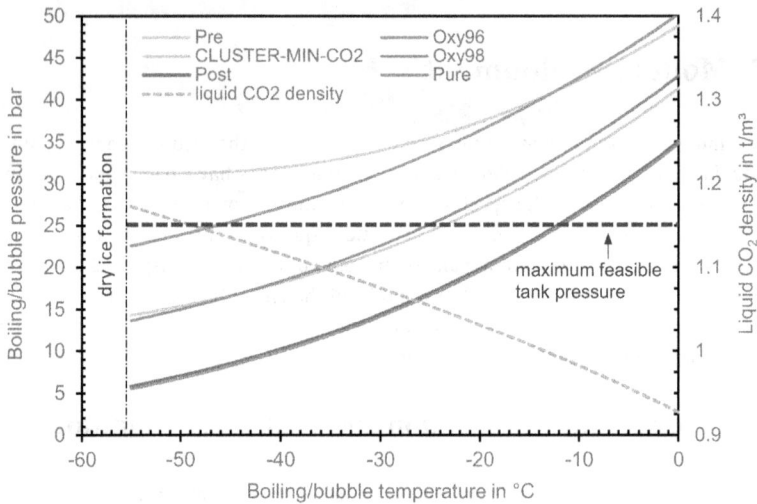

Figure 14: Boiling/bubble pressure for typical CO_2 stream compositions as well as the density of pure CO_2

The CO_2 stream compositions are given in Table 5. These stream compositions are based on the COORAL project [27] and have been slightly adapted to reduce the water concentration from 100 ppmv to 50 ppmv. The lower water concentration of 50 ppmv has been proposed in more recent literature to avoid corrosion [65]. The Post-Combustion CO_2 stream (Post) is based on a typical CO_2 stream composition for amine based processes [23, 66, 67]. The Oxyfuel CO_2 streams (Oxy98 and Oxy96) are based on typical values for a 2-stage partial condensation gas processing unit (GPU) [66, 68, 69]. The Pre-Combustion CO_2 stream composition (Pre) is based on the Selexol™ process [21]. The CL-MIN-CO_2 CO_2 stream results from a cluster of different CO_2 emitters which feed into a central pipeline infrastructure. The cluster consists of three lignite-fired power plants (with Pre-Combustion, Post-Combustion and Oxyfuel CO_2 capture, respectively), two hard-coal-fired power plants (Post-Combustion and Oxyfuel CO_2 capture), two gas-combined-cycle power plants (Post-Combustion CO_2 capture) as well as a blast-furnace steel plant and a cement plant with Post-Combustion CO_2 capture. Each of these plants is operated according to its typical load characteristics, e.g. the steel plant is operated at constant load while the two combined-cycle gas plants are used for peak power demand. Consequently, the CO_2 stream composition in the pipeline is variable. The different CO_2 sources and the pipeline infrastructure are discussed in detail in another dissertation [70]. CL-MIN-CO_2 is the composition with the highest amount of impurities that occurs in the central pipeline during the considered reference year 2016.

Table 5: **Typical CO_2 and impurity concentrations in vol.-% for Post-Combustion, Oxyfuel and Pre-Combustion CO_2 capture processes**

Component in vol.-%	Post	Oxy98	Oxy96	Pre	CL-MIN-CO_2
CO_2	99.93	98.003	96.563	98.005	98.653
N_2	0.023	0.710	1.960	0.900	0.513
O_2	0.015	0.670	0.810		0.322
Ar	0.023	0.590	0.570	0.030	0.292
H_2O	0.005	0.005	0.005	0.005	0.005
NO_x	0.002	0.010	0.010		0.008
SO_x	0.001	0.007	0.007		0.004
CO	0.001	0.005	0.075	0.040	0.010
H_2				1.000	0.189
H_2S				0.005	0.001
COS				0.005	0.001
CH_4				0.010	0.002

Figure 14 shows that the bubble pressure generally increases with an increased fraction of impurities. A maximum transport pressure of 25 bar is assumed for the techno-economic feasibility of ship transport. This pressure limit is deduced from the study of the mechanical design calculations for CO_2 tanks presented in section 2.3. With a bubble pressure of 25 bar, the design pressure for the CO_2 will most likely be between 26 bar and 27 bar, depending the maximum combined internal liquid pressure $(p_{gd})_{max}$. For a tank length of 50 m, the resulting tank volume would be in the range of approximately 400 m^3 to 900 m^3, depending on the allowable membrane stress. If the design pressure is halved to 13 bar, the tank volume would be approximately four times greater, with a range of 1800 m^3 to 3600 m^3. Thus, a low bubble pressure is preferable. For the Pre CO_2 stream, which contains high concentrations of volatile components (e.g. H_2, N_2, O_2), the bubble pressure is generally above 25 bar. This means that this stream cannot be transported by ship

31

when considering the imposed pressure limit. For the Oxy96 CO_2 stream, the bubble pressure is below 25 bar for temperatures up to approximately -43 °C. For the Oxy98 CO_2 stream, temperatures up to approximately -25 °C are within the imposed pressure limit. For the Post and the pure CO_2 streams, a temperature of up to approximately -12 °C is feasible.

In this work, a transport temperature of -50 °C is assumed. This value is in line with most literature on large-scale CO_2 ship transport [6–8, 34, 38, 40, 41, 71]. With the exception of the Pre CO_2 stream, all CO_2 streams considered can be transported at this temperature level as the bubble pressures are below 25 bar. For the Oxy98, the Post and the pure CO_2 stream, a higher transport temperature would also be possible. This would lead to a lower energy demand for liquefaction and injection but, on the other hand, to increased investment costs for tank construction due to a higher bubble pressure. A higher bubble pressure would require increased wall thickness and a higher temperature is associated with a higher tank volume per quantity of CO_2 due to decreased density. A higher pressure would also lead to a higher number of tanks, as the maximum tank wall thickness is usually limited to 50 mm for typical low temperature steels (q. v. section 2.3). For this reason, lower transport temperature is generally considered to be more economical than high transport pressure [24, 31, 72, 73]. A side benefit of a lower CO_2 temperature is increased density. The temperature dependency of liquid CO_2 density is shown in Figure 14. As for most liquids, the compressibility of liquid CO_2 can be neglected so that the density in the liquid regime is mainly determined by temperature. In a previous work it has been shown that the impact of impurities on the liquid density is small for the considered CO_2 streams [74]. Generally, the liquid density slightly decreases with increased impurity concentrations.

3.2 CO_2 Liquefaction Process

The purpose of modelling various liquefaction processes is to determine the minimum specific energy demand for CO_2 liquefaction. It is defined as the sum of the minimum electrical energy input required for liquefaction. The term 'minimum' is used to indicate that sensitivity analyses were carried out to determine the optimum operating parameters for each process. Secondly, the term indicates that the energy demand of the actual process would be slightly higher as minor effects like the electrical demand of the cooling water pump or the pressure drops of the heat exchangers have been neglected. The minimum specific energy demand is defined more precisely in the following.

Table 6:   Boundary conditions for the simulations

Parameter	
CO2 input stream conditions	15 °C, 100 bar
CO2 output stream conditions	-50 °C, pressure depending on impurity concentrations
Ambient temperature	20 °C
Cooling water (seawater) temperature	15 °C
Minimum internal temperature approach seawater heat exchanger	5 K
Minimum internal temperature approach internal heat exchanger	3 K
Compressor efficiency	85 % polytropic 97 % electromechanical
Superheat at compressor inlet	5 K
Liquid CO2 expander hydraulic efficiency	90 %
Two-phase CO2 expander isentropic efficiency	70 %

The boundary conditions for all simulations are shown in Table 6. A cooling water temperature of 15 °C was assumed based on average coastal seawater temperatures in the Northern part of central Europe. For large-scale seawater cooling systems, a minimum internal temperature approach of 3-5 K is attainable [75]. For the seawater heat exchangers in this work, a minimum internal temperature approach of 5 K is used as a conservative estimate. For the internal heat exchangers, a minimum internal temperature approach of 3 K is assumed , which is a typical value from literature [8, 39]. The electromechanical efficiency of 97 % is a combination of the electrical motor efficiency of 98 % (synchronous motor in MW range) and a mechanical efficiency of 99 %. The polytropic efficiency of the compressors is assumed to be 85 % [7]. All simulations were carried out with Aspen Plus© V8.6 using the Peng-Robinson equations of state in a modification by Boston and Mathias [76]. This equation of state has provided satisfactory results for the vapour-liquid equilibria and critical points of CO_2-rich mixtures [77–79]. However, it is generally recognised that there are still significant knowledge gaps and inaccuracies in regard to experimental data and theoretical models for the thermodynamic properties of CO_2-rich mixtures [77, 80–82].

3.2.1 Closed Cycle Process

The 1-stage and the 2-stage closed cycle process presented in Chapter 2 (Figure 3 and Figure 4) as well as a 3-stage closed cycle process have been modelled to determine the impact of the refrigerant selection on the specific electrical energy demand of the refrigerant compressors. The results of this analysis are presented in Chapter 4.

To reduce the specific energy demand of the process, several improvements for the closed cycle process have been analysed. The 1-stage, the 2-stage and the 3-stage closed cycle base process are used as references for the evaluation. Figure 15 shows the 2-stage closed cycle base process. All streams and components have been labelled to simplify the description of the process. The results for pressure, temperature and vapour fraction apply for pure CO_2 with ammonia as the refrigerant. These results are discussed in Chapter 4. Contrary to the processes presented in Chapter 2 (Figure 3 and Figure 4), it has been taken into account that the refrigerant is usually superheated before entering the compressor. For this purpose, additional heat exchangers were added which use the CO_2 stream before expansion (i.e. prior temperature reduction) as the heating medium (PH1 and PH2 in Figure 15).

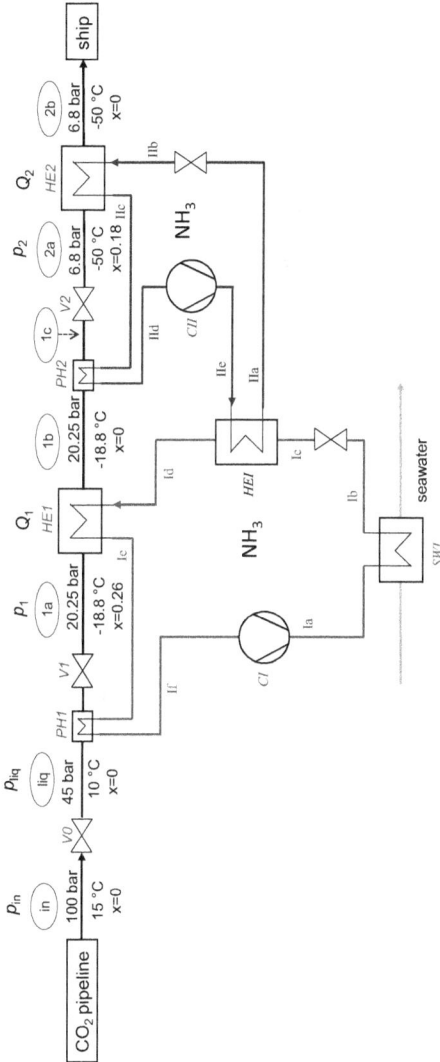

Figure 15: 2-stage closed cycle base process with refrigerant superheating before compression (*PH1, PH2*). Pressures, temperatures and vapour fractions apply to pure CO_2.

35

In the 2-stage closed cycle base process (Figure 16), the CO_2 stream pressure is reduced from p_{in} to p_{liq}, i.e. from pipeline pressure in supercritical state (100 bar) to liquid state (45 bar). p_{liq} is the lowest pressure at which no vapour is formed, hence it is equal to the boiling pressure. The expansion from p_{in} to p_{liq} is merely done for comparison with the optimised closed cycle process, in which the expansion to p_{liq} is used for energy recovery. It is an isenthalpic expansion and therefore associated with a temperature decrease (5 K). In heat exchanger *PH1*, the temperature of the CO_2 stream is slightly reduced (by approximately 0.2 K in the case of pure CO_2 in combination with NH_3) to attain a 5 K superheat for the refrigerant in refrigeration stage I. Next, the CO_2 stream is expanded to p_1 into the two-phase regime (20.25 bar). The vaporous fraction of the stream is condensed in *HE1*. In practice, a phase separator would probably be installed upstream of *HE1* so that only vaporous CO_2 enters the heat exchanger and the liquid phase would be sent to a bypass line. This is not shown in Figure 15 for reasons of simplification. Next, the CO_2 stream is slightly subcooled in *PH2* to provide the heat duty necessary for superheating the refrigerant in cycle II. Subsequently, the CO_2 stream is expanded to p_2 (6.8 bar) and then condensed in *HE2*. Likewise, the CO_2 stream would in practice probably be separated into liquid and vaporous phases prior to entering the heat exchanger. At position *2b*, the CO_2 stream is in liquefied state at the pressure and temperature levels required for ship transport.

Two vapour-compression refrigeration cycles in cascade configuration are utilised - stage I and stage II. The refrigerant is superheated in *PH1* (stage I) and *PH2* (stage II) and then compressed in *CI* and *CII,* respectively. In stage I, the pressure after compression (p_{Ia}) is determined by the seawater temperature and the minimum internal temperature approach in *SWI*. In stage II, the pressure at this position (p_{IIe}) is determined by the CO_2 pressure p_1: The pressure p_1 determines the corresponding CO_2 temperature T_{1a} and - in combination with the minimum internal temperature approach of the CO_2 condenser *HE1* - the refrigerant temperatures and pressures of streams 'Ic', 'Id', and 'Ie'. Together with the minimum internal temperature approach in *HEI*, the temperatures and pressures of streams 'Ic' and 'Id' determine the pressure and temperature in stage II after compression (T_{IIe} and p_{IIe}). The condensation of the refrigerant is carried out in *SWI* (stage I) and *HEI* (stage II). Subsequently, the refrigerant is expanded into the two-phase regime to reduce the temperature. The cooling duty for the CO_2 stream is provided by the evaporation of the refrigerant in *HE1* or *HE2*, respectively. The log p-h diagram of the 2-stage closed cycle base process is shown in Figure 16. The points are labelled according to Figure 15. The minimum specific energy demand of the 2-stage closed cycle process, $P_{min,spec}$, is defined as:

$$P_{min,spec} = \frac{P_{CI} + P_{CII}}{\dot{m}_{CO2}} \qquad (3.1)$$

P_{CI} is the energy demand of compressor CI, P_{CII} the energy demand of compressor CII and $\dot{m}_{CO2}$ the mass flow rate of the CO_2 stream.

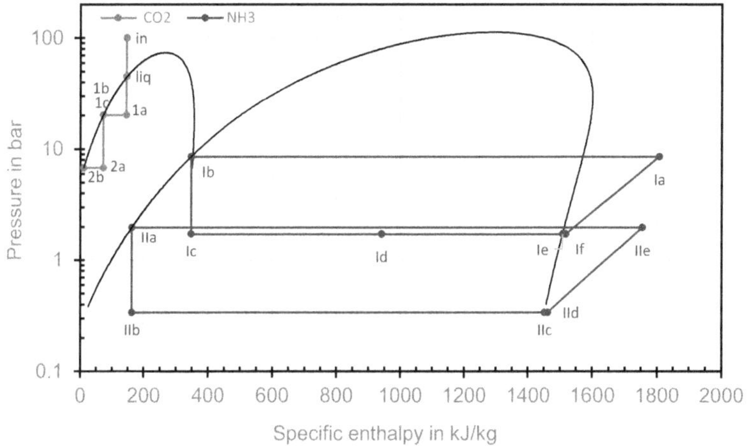

Figure 16: Log p-h diagram of the 2-stage closed cycle base process. The points are labelled according to Figure 15.

Figure 17 shows the 3-stage closed cycle base process. Different to the 2-stage closed cycle process, a third refrigeration cycle, Stage III, and a third expansion stage for the CO_2 stream have been added. The log p-h diagram of the 3-stage closed cycle base process is shown in Figure 18. The minimum specific energy demand of the 3-stage closed cycle process is defined analogically to eq. (3.1):

$$P_{min,spec} = \frac{P_{CI} + P_{CII} + P_{CIII}}{\dot{m}_{CO2}} \qquad (3.2)$$

37

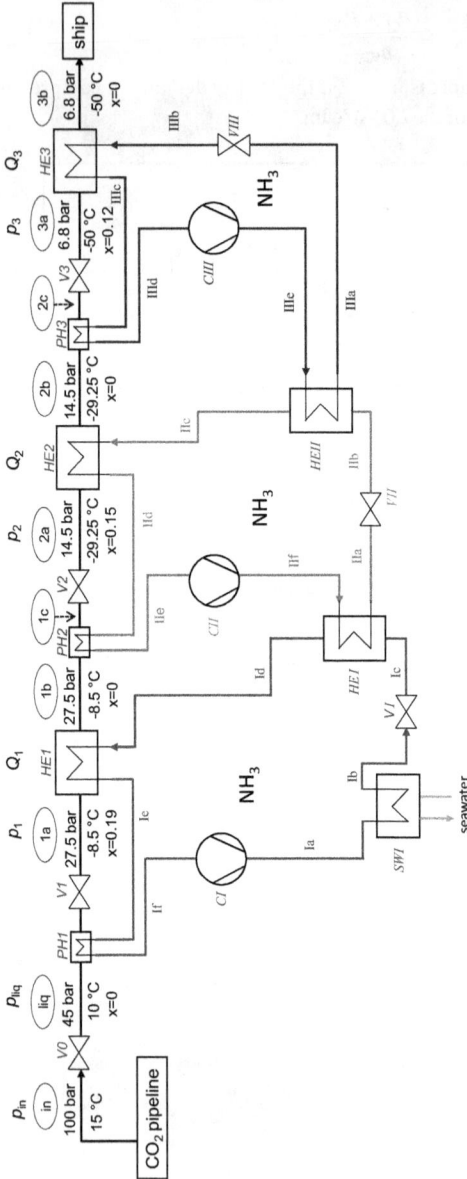

Figure 17: 3-stage closed cycle base process with refrigerant superheating before compression (*PH1*, *PH2*, *PH3*). Pressures, temperatures and vapour fractions apply to pure CO_2.

38

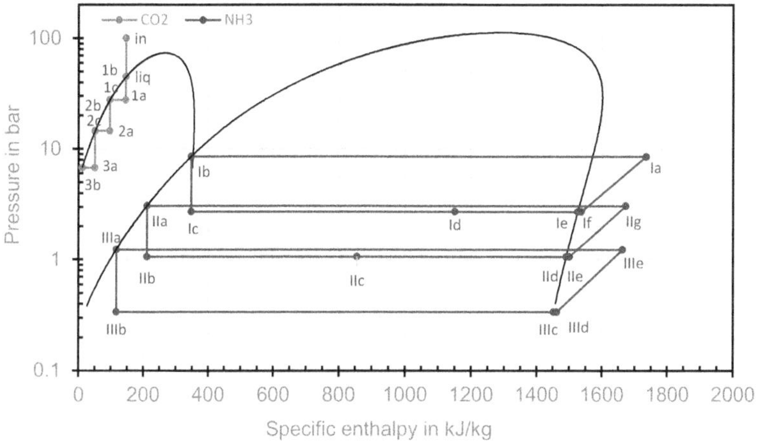

Figure 18: Log p-h diagram of the 3-stage closed cycle base process. The points are labelled according to Figure 17.

Five measures of improvement are evaluated for the closed cycle processes – two regarding the CO_2 stream and three regarding the refrigeration cycles. These five improvements – a liquid expander, a two-phase expander, a phase separator, an aftercooler and an internal heat exchanger – are shown in Figure 19 and Figure 20 for the 2-stage and the 3-stage closed cycle, respectively. The temperatures and pressures in these figures are based on the simulation results for a pure CO_2 stream with ammonia as the refrigerant. They will be discussed in Chapter 4.

The general idea of the improvements to the CO_2 stream side is to recover some of the mechanical energy during depressurisation by using expanders instead of valves. A side benefit of extracting mechanical energy from the CO_2 stream is that the temperature is lowered due to the reduction of enthalpy, thus a lower heat duty is required for liquefaction. Generally, expanders can operate both in the liquid and in the two-phase (liquid and vaporous) regime. Cryogenic hydraulic turbines - often termed liquefied gas expanders - are widely used in the LNG industry and high efficiencies of around 90 % are attainable [83]. Hence, it is assumed that a similar liquid expander could be developed for CO_2. In the optimised closed cycle processes, the liquid expander $EX0$ is used to expand the CO_2 stream from p_{in} to p_{liq}. As mentioned earlier, p_{liq} is equal to the boiling pressure, i.e. the lowest pressure possible for expansion into the liquid phase regime.

39

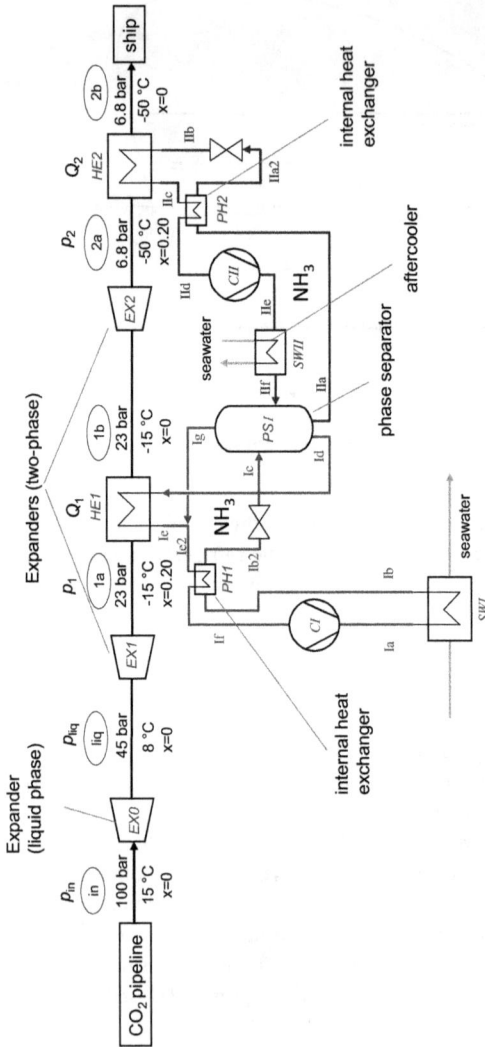

Figure 19: Optimised 2-stage closed cycle process with the five indicated improvements. Temperatures, pressures and vapour fractions apply to pure CO_2.

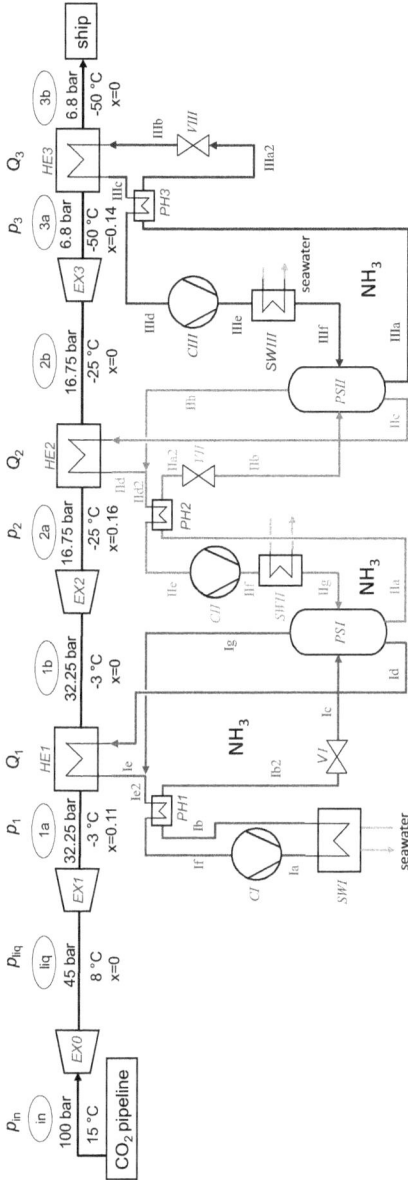

Figure 20: Optimised 3-stage closed cycle process with the five improvements liquid CO_2 expander (EX0), two-phase CO_2 expanders (EX1, EX2, EX3), phase separators (PSI, PSII), aftercoolers (PSI, PSII), aftercoolers (SWII, SWIII) and internal heat exchangers (PH1, PH2). Temperatures, pressures and vapour fractions apply to pure CO_2.

In contrast to expanders operating in liquid regime, two phase expanders for liquefied gases are not yet an established technology. For the LNG industry, there is already one product commercially available [84, 85]. This two-phase LNG expander is designed as a radial outflow turbine with a draft tube and installed in upright position so that the direction of flow matches the natural, upward draft of the vapour [86]. In this work it is assumed that a similar two-phase expander could also be developed for CO_2. In the optimised 2-stage closed cycle process, two two-phase expanders (*EX1* and *EX2*) are used, while three two-phase expanders (*EX1, EX2* and *EX3*) are used in the optimised 3-stage closed cycle process. The efficiency of two-phase expanders is generally lower than the efficiency of liquid phase expanders (i.e. hydraulic turbines) - a conservative estimate of 70 % isentropic efficiency is used in this work. The log p-h diagram of the optimised 2-stage closed cycle process is shown in Figure 21 and the log p-h diagram of the optimised 3-stage closed cycle process in Figure 22. The impact of the CO_2 expanders is reflected in the decreasing enthalpy during expansion. In comparison, the enthalpy during expansion is constant in the closed cycle base processes (Figure 16 and Figure 18) as valves are used.

For the refrigeration cycles, the three measures of optimisation are a compressor aftercooler, the replacement of the cascade heat exchanger by a phase separator, and the replacement of the refrigerant superheater by an internal heat exchanger. The purpose of the aftercooler is to reduce the inlet temperature of the cascade heat exchanger. With ammonia as the refrigerant, the temperature after compression is 98 °C in the lower temperature cycle (stage II) of the 2-stage closed cycle base process. In the 3-stage closed cycle base process, the temperatures are 47 °C (stage II) and 92 °C (stage III). By installing an aftercooler, the inlet temperature of the cascade heat exchanger can be reduced to 20 °C (in the reference case), assuming seawater as the cooling medium. The benefit is that less heat is transferred in the cascade heat exchanger and the refrigerant mass flow rate in the upper temperature refrigeration stage(s) is reduced. One aftercooler is used for the optimised 2-stage closed cycle process (*SWII* in Figure 19) and two aftercoolers are used for the optimised 3-stage closed cycle process (*SWII* and *SWIII* in Figure 20). The impact of the aftercoolers is reflected in the log p –h diagram, by point 'IIf' in Figure 21 and by points 'IIg' and 'IIIf' in Figure 22. It can be seen that the heat duty transferred in the cascade heat exchangers is significantly reduced due to the aftercoolers.

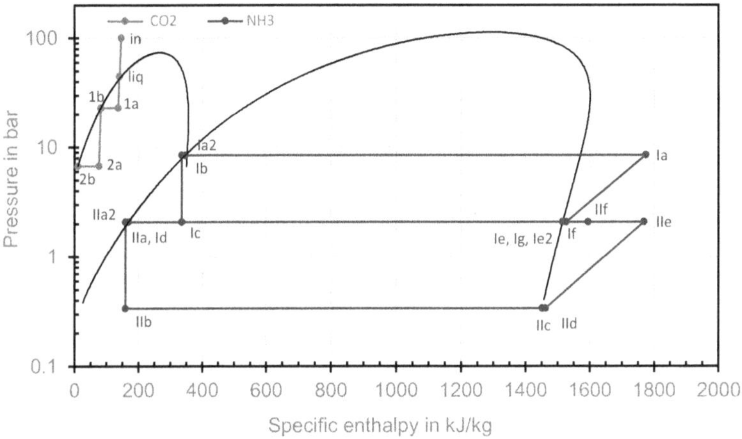

Figure 21: Log p-h diagram of the optimised 2-stage closed cycle process. The points are labelled according to Figure 19.

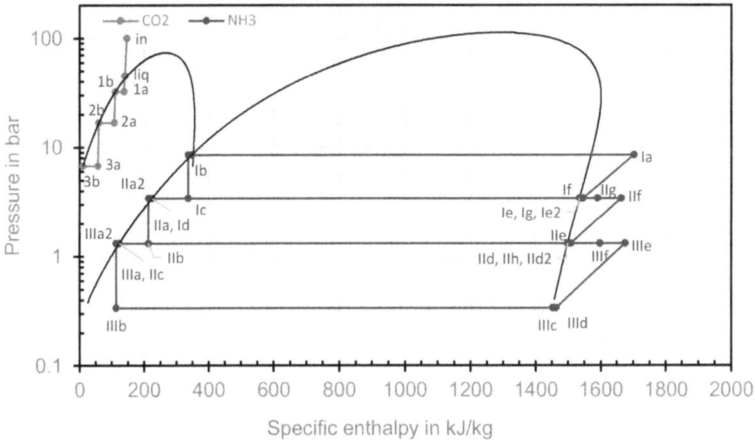

Figure 22: Log p-h diagram of the optimised 3-stage closed cycle process. The points are labelled according to Figure 20.

The second measure of optimisation for the refrigerant cycle is the replacement of the cascade heat exchanger by a phase separator. As the same refrigerant type is used in all refrigeration cycles, the upper and lower cycle refrigerants can be mixed. Therefore, a phase separator can fulfil the same purpose as the cascade heat exchanger, which is the condensation of the lower temperature cycle refrigerant by evaporation of the upper cycle one. In the optimised 2-stage closed cycle process, one phase separator is used (*PSI*), in the optimised 3-stage closed cycle process, two phase separators are used (*PSI* and *PSII*). The working principle of the phase separator is explained by the example of *PSI* in the 2-stage closed cycle process: Gaseous refrigerant (stream 'IIf' in Figure 19) enters from the lower temperature refrigeration cycle and is mixed with the two-phase refrigerant from the upper temperature cycle (stream 'Ic'). The liquid refrigerant fraction is partially sent to the lower temperature refrigeration cycle expansion valve (stream 'IIa') and partially used for the CO_2 condenser in the upper temperature cycle (stream 'Id'). The vaporous fraction (stream 'Ig') is mixed with the vaporous refrigerant stream 'Ie' from the CO_2 condenser *HE1*. The resulting stream 'Ie2' is superheated in *PH1* and sent to the upper temperature cycle compressor *CI*. The benefit of using a phase separator instead of a heat exchanger is that a temperature difference between the lower temperature cycle condensation and the upper temperature cycle evaporation temperature is not necessary. If a cascade heat exchanger were used instead, the lower cycle condensation temperature would need to be higher than the upper cycle evaporation temperature, thus, a higher output pressure for the lower cycle compressor *CII* would be necessary. Therefore, the energy demand is lower when the cascade heat exchanger is replaced by a phase separator. In the log p-h diagrams (Figure 21 and Figure 22), the impact of the phase separator is reflected by the overlap of the lower temperature refrigeration cycle condensation line and the upper cycle evaporating line. Contrary to the optimised closed cycle processes, there is a visible gap between these lines in the log p-h diagrams of the base processes (Figure 16 and Figure 18).

The third measure of optimisation for the refrigeration cycle is the replacement of the refrigerant superheater by an internal heat exchanger. In the optimised 2-stage closed cycle process, two internal heat exchangers are used (*PHI* and *PHII* in Figure 19) and in the optimised 3-stage closed cycle process, three internal heat exchangers are used (*PHI, PHII* and *PHIII* in Figure 20). The internal heat exchangers achieve the required superheat at the compressor inlet by transferring heat from the high pressure, high temperature condensed refrigerant to the low pressure evaporated refrigerant of the same cycle. Thus, they serve the same purpose as *PHI* and *PHII* in the 2-stage closed cycle base process (Figure 15) and *PHI, PHII* and *PHIII* in the 3-stage closed cycle base process (Figure 17). The only difference is that the refrigerant instead of the CO_2 stream is used as a heat source, which leads to subcooling of the refrigerant prior to expansion. Subcooling of the refrigerant prior to expansion, in turn, leads to a lower vapour fraction after expansion, which reduces the refrigerant mass flow rate necessary to transfer the desired heat duty in the refrigerant evaporator. In the case of the 2-stage closed cycle process, the refrigerant in cycle II is subcooled by 2.2 K before expansion (stream 'IIa2') which leads to a reduction in the vapour fraction from 10.4 % to 10.2 % after expansion (stream 'IIb').

3.2.2 Open Cycle Process

The 1-stage open cycle base process introduced in Chapter 2 and a 3-stage open cycle process have been modelled for comparison with the respective closed cycle processes. The idea of the 3-stage open cycle process is to replace the single compressor with a high pressure ratio ($\Pi = \frac{57.4}{6.75} \approx 8.5$) by three compressors, each with a pressure ratio of approximately 2. This idea is based on the work of Yoo et al. [34]. The 3-stage open cycle process is shown in Figure 23.

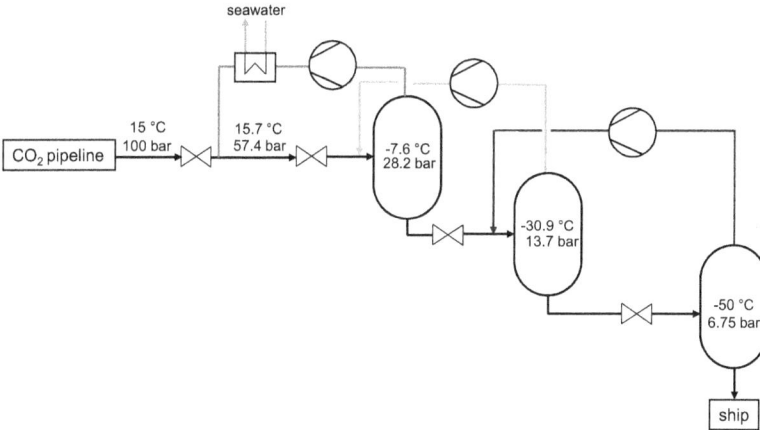

Figure 23:   3-stage open cycle process for CO_2 liquefaction. The temperatures and pressures apply for pure CO_2.

45

A major disadvantage of the open cycle process is the fact that it only works with a pure CO_2 feed stream. During the simulations it was possible to obtain a steady-state solution for pure CO_2, but not for CO_2 streams with impurities. Further analyses showed that this is a conceptual problem with the open cycle process rather than a numerical or modelling error. This can be illustrated when considering the mass balance shown in Figure 24, which is a simplified representation of the 1-stage open cycle process: The CO_2-rich feed stream $\dot{m}_p$ with a CO_2 concentration of x^p_{CO2} is entering the system boundary. Only one stream is leaving the system boundary, which is $\dot{m}_s$ with a CO_2 concentration of x^s_{CO2}. A mass balance shows that the input and output mass flow rates and CO_2 concentrations must be equal in steady state:

$$\dot{m}_p = \dot{m}_s \text{ and } x^p_{CO2} = x^s_{CO2} \tag{3.3}$$

System boundary

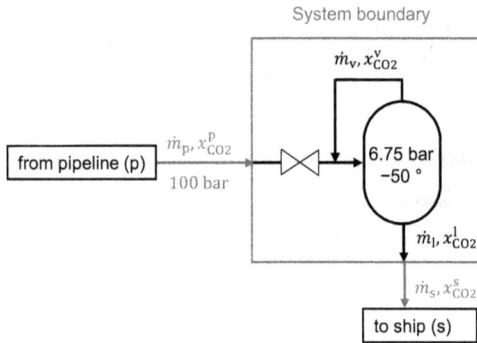

Figure 24: **Mass balance of the 1-stage open cycle process shown in Figure 6**

At the same time, the conditions of the phase separator are determined by the desired transport conditions, e.g. 6.75 bar and -50 °C. If these conditions result in a phase separation into $\dot{m}_v$ and $\dot{m}_l$, the liquid fraction is equal to the output mass flow rate: $\dot{m}_l = \dot{m}_s$. On the other hand, any CO_2-rich mixture in vapour-liquid equilibrium with volatile components (e.g. N_2, O_2, H_2) has a higher concentration of these components in the vaporous phase so that

$$x^v_{CO2} < x^l_{CO2} = x^s_{CO2} \tag{3.4}$$

At the same time, eq. (3.3) must be fulfilled. This means that

$$x^p_{CO2} = x^s_{CO2} = x^l_{CO2} \quad \text{and } \dot{m}_p = \dot{m}_s = \dot{m}_l \qquad (3.5)$$

which is only possible when $\dot{m}_v = 0$. However, this would render the process useless, as its concept is to use the recycle stream $\dot{m}_v$ for refrigeration. Thus, the open cycle process shown in Figure 6 is not feasible for CO_2 with impurities.

Purification of the recycle stream would be necessary to make this process feasible, so that $x^p_{CO2} = x^s_{CO2} = x^v_{CO2}$. Otherwise, volatile components would accumulate. In the context of this work, purification is not considered to be a viable option as the CO_2 feed stream originates from a pipeline and has already been purified at the respective CO_2 capture plant. This is significantly different to the original open cycle process by Aspelund and Jordal [9] in which a low pressure CO_2 feed stream from a capture plant is assumed. In that case, purification is necessary and thus, it is included in the original open cycle process. Due to the different CO_2 feed stream conditions in this work, the open cycle process will not be further studied for CO_2 streams with impurities. However, it will be used as a benchmark process for the liquefaction of pure CO_2 with the closed cycle process.

3.3 Boil-Off Gas

3.3.1 Boil-Off Gas Formation

The amount of boil-off gas can be determined by calculating the heat input into the tanks, $\dot{Q}_{BOG}$, and dividing it by the enthalpy of evaporation, ΔH_v, of the CO_2:

$$\dot{m}_{BOG} = \frac{\dot{Q}_{BOG}}{\Delta H_v} \qquad (3.6)$$

The enthalpy of evaporation ΔH_v depends on the pressure and the composition of the CO_2 stream. Values for the enthalpy of evaporation of the analysed CO_2 streams are presented in section 4.2. The heat input $\dot{Q}_{BOG}$ is calculated by using the conventional heat transfer equation. The total heat input for a ship with a certain number of tanks, n_{tanks}, is given by:

$$\dot{Q}_{BOG} = n_{tank} \cdot U_{tank} \cdot A_{tank} \cdot (T_{tank} - T_{air}) \qquad (3.7)$$

A tank temperature T_{tank} of -50 °C and an ambient temperature T_{air} of 15 °C are assumed. For simplification, cylindrical pipe wall geometry is assumed for the tanks so that the overall heat transfer coefficient and heat transfer area can be determined by summing up the overall heat transfer coefficients of the longitudinal pipe wall ($U_{\text{wall}} \cdot A_{\text{wall}}$) and the two cylinder heads $U_{\text{heads}} \cdot A_{\text{heads}}$:

$$U_{\text{tank}} \cdot A_{\text{tank}} = U_{\text{wall}} \cdot A_{\text{wall}} + U_{\text{heads}} \cdot A_{\text{heads}} \tag{3.8}$$

The overall heat transfer coefficient and heat transfer area of the longitudinal pipe wall is calculated by [87]:

$$
U_{\text{wall}} \cdot A_{\text{wall}} = \pi \cdot l_{\text{tank}} \cdot \left[\frac{1}{h_{\text{CO2}} \cdot d_i} + \frac{1}{\lambda_{\text{steel}}} \cdot \ln\left(\frac{d_i + 2t_{\text{steel}}}{d_i}\right) \right.
$$
$$
\left. + \frac{1}{\lambda_{\text{ins}}} \cdot \ln\left(\frac{d_i + 2(t_{\text{steel}} + t_{\text{ins}})}{d_i + t_{\text{steel}}}\right) + \frac{1}{h_{\text{air}} \cdot d_o} \right]^{-1} \tag{3.9}
$$

The overall heat transfer coefficient and heat transfer area of the two tank heads is estimated by:

$$
U_{\text{heads}} \cdot A_{\text{heads}} = \frac{\pi d_o^2}{2 \dfrac{1}{h_{\text{CO2}} \cdot d_i^2/d_o^2} + \dfrac{t_{\text{ins}}}{\lambda_{\text{ins}}} + \dfrac{t_{\text{steel}}}{\lambda_{\text{steel}}} + \dfrac{1}{h_{\text{air}}}} \tag{3.10}
$$

The convective heat transfer coefficients are h_{CO2} for the inner surface of the tank and h_{air} for the outer surface. The thermal conductivities of the tank wall are λ_{steel} for the steel layer and λ_{ins} for the insulation layer. l_{tank} is the tank length, d_i the inner diameter and d_o the outer diameter of the tank. The inner diameter of the tank is calculated from the design pressure, the allowable stress and the wall thickness according to eq. (2.3). The values for all relevant parameters are shown in Table 7. For the steel shell, a wall thickness of 50 mm is assumed. Polyurethane foam with a thickness of 15 cm is used for insulation. All material properties are based on literature values [50].

Table 7:   Boundary conditions for the boil-off gas calculations

Parameter	Symbol	Unit	Value
Convective heat transfer coefficient liquid CO_2	h_{CO2}	$W/(m^2K)$	100
Wall thickness steel	t_{steel}	m	0.05
Thermal conductivity steel	λ_{steel}	W/mK	25
Thickness of insulation	t_{ins}	m	0.15
Thermal conductivity insulation (PU foam)	λ_{ins}	W/mK	0.02
Convective heat transfer coefficient air	h_{air}	$W/(m^2K)$	10
Ambient air temperature	T_{air}	°C	15

3.3.2 Boil-Off Gas Reliquefaction

For the reliquefaction of boil-off gas, the 2-stage and the 3-stage closed cycle processes are studied. The 2-stage closed cycle boil-off gas reliquefaction process for pure CO_2 is shown in Figure 25. Unlike in the closed cycle process for the liquefaction of a pipeline CO_2 stream, the feed stream is fully evaporated, at tank temperature (-50 °C), and at tank pressure (6.8 bar for pure CO_2). Thus, the heat transfer from the CO_2 stream to the refrigeration cycle is carried out at -50 °C only. The boundary conditions shown in Table 6 are assumed.

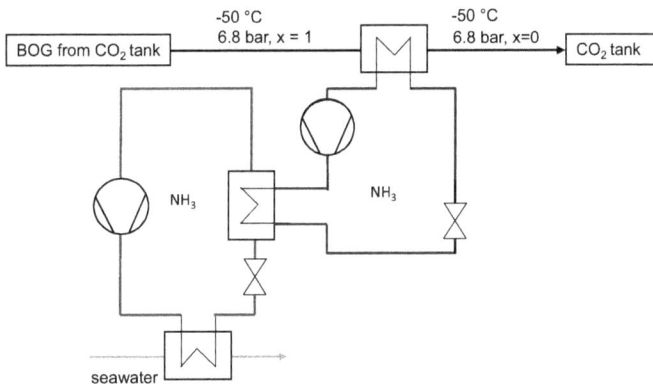

Figure 25:   2-stage closed cycle boil-off gas reliquefaction process. Temperatures, pressures and vapour fractions apply to pure CO_2.

49

3.4 Injection Process

The pressure must be raised from tank pressure on ship to wellhead pressure for the injection. Additionally, the CO_2 stream temperature must be raised to a temperature above 0 °C before injection to avoid hydrate formation. Figure 26 shows the model of the injection process with the integration of the injection pump, the engine and the available heat sources. The calculation of the overall energy demand for injection works as follows: First, the electrical energy demand of the injection pump is determined in dependency of the inlet pressure (i.e. tank pressure) and outlet pressure (i.e. wellhead pressure). From the electrical energy demand of the injection pump, the mechanical shaft power of the engine can be calculated. When the shaft power is known, the waste heat streams of the engine can be calculated. The thermal energy demand for the heating of the CO_2 is provided by seawater and by engine waste heat. The heat stream of the seawater depends on seawater temperature, and the heat stream of the engine depends on the inlet and outlet pressure of the injection pump. For low seawater temperatures and wellhead pressures, additional heat might be necessary to reach the desired CO_2 temperature of 5 °C. A temperature of 5 °C is thought to be sufficient to preclude both hydrate formation and the build-up of ice at the outer surfaces of the injection hose [15].

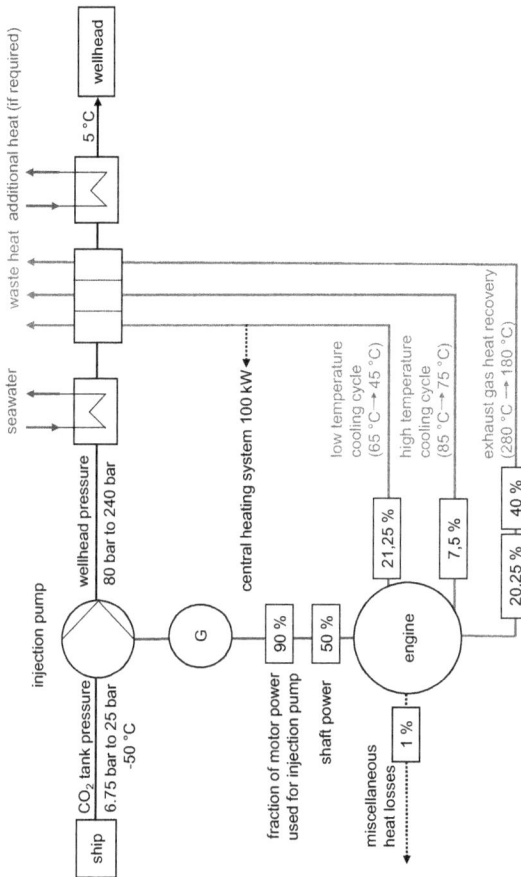

Figure 26: Integration of injection pump, engine and different heat sources for the injection of CO_2

As illustrated in Figure 26, a hydro-mechanical efficiency of 70 % for the injection pump is assumed. This value is in line with literature values for low temperature CO_2 pumps [18, 78]. A medium speed diesel engine with an overall efficiency of 50 % is connected to the injection pump via a generator. The engine also provides auxiliary energy for the various components of the ship (seawater pumps, lubrication system, ventilation, etc., hotel function for staff), so that only 90 % of the engine power is used for the injection pump. Engine waste heat is rejected via a high temperature and a low temperature cooling cycle as well as the exhaust gas stream. The distribution of the energy streams and their individual temperatures are derived from literature [88, 89]. The engine's high temperature cooling cycle is used to cool highly stressed engine parts such as the cylinders and cylinder heads. Only a small temperature difference between inlet and outlet temperature is acceptable (e.g. 75 °C inlet and 85 °C outlet temperature of the cooling medium in the engine's cooling cycle heat exchanger). A heat stream of 100 kW is assumed to be required for the central heating cycle of the ship, hence it is subtracted from the high temperature cooling cycle heat stream that is available for CO_2 injection.

The engine's low temperature cooling cycle is used to cool other parts of the engine such as the lubrication oil and the charge air intercooler. The heat from the low temperature cycle is usually rejected to seawater. In the model, it is used for CO_2 heating, as illustrated in Figure 26. Another waste heat source on ships is exhaust gas. Commercial systems for exhaust gas heat recovery already exist, so that their use for CO_2 injection is probable. The exhaust gas exit temperature is assumed to be 180 °C [90]. Therefore, only approximately 40 % of the exhaust gas energy can be recovered. A heat transfer oil cycle is utilised to avoid direct heat transfer to the cold CO_2 stream and the associated formation of sulphuric acid.

3.5 Transport Chain Model

A model of the transport chain encompassing liquefaction, onshore intermediate storage, loading, transport and injection has been developed. This model is used for the dimensioning of the individual transport chain components and applied to several scenarios. The results of the transport chain dimensioning analysis are used to calculate the energy demand for liquefaction, boil-off gas reliquefaction and injection, assuming the respective specific energy demands determined in this work. While the energy demand for liquefaction could be calculated directly from the CO_2 source mass flow rate, i.e. without a transport chain model, the injection mass flow rate, and therefore the energy demand of the injection, is significantly influenced by the parameters of the transport chain. The transport chain model provides a preliminary estimation for the dimensioning of the components and the energy demand – certain aspects that would be considered in a detailed engineering design are neglected, such as safety margins or the frequency of storms during which ship transport is suspended.

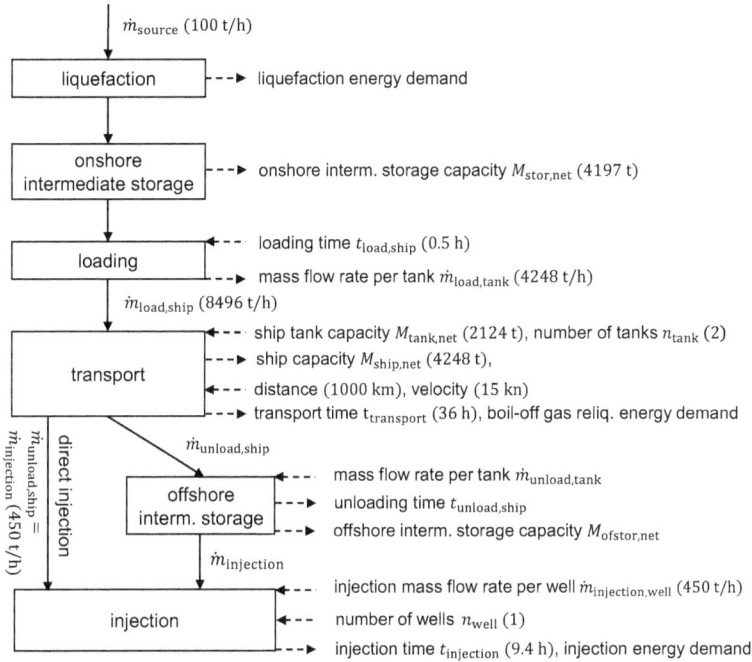

Figure 27: Individual components and mass flow rates of the CO_2 transport chain model. The values in parentheses represent an exemplary transport chain.

Figure 27 shows the components of the implemented transport chain model with the most important input and output parameters shown on the right. The values in parentheses represent an exemplary transport chain with two ships, a source mass flow rate $\dot{m}_{source}$ of 100 t/h, a transport distance of 1000 km and a transport speed of 15 kn. The loading mass flow rate $\dot{m}_{load,ship}$ was selected so that it corresponds with a loading time $t_{load,ship}$ of 0.5 h. The CO_2 is directly injected from the ship with a mass flow rate of 450 t/h (direct injection), which means that the unloading mass flow rate $\dot{m}_{unload,ship}$ is equal to the injection mass flow rate $\dot{m}_{injection}$. The roundtrip time of a ship $t_{roundtrip}$ consists of the loading time $t_{load,ship}$, the transport time $t_{transport}$ (counted twice) and the unloading time $t_{unload,ship}$. Additionally, a certain time period for handling (manoeuvring, mooring etc.) is assumed before and after loading ($t_{handling,load}$) and unloading ($t_{handling,unload}$):

$$t_{\text{roundtrip}} = t_{\text{load,ship}} + t_{\text{unload,ship}} + 2(t_{\text{handling,load}} \\ + t_{\text{transport}} + t_{\text{handling,unload}}) \tag{3.11}$$

For the exemplary values shown in Figure 27, handling times of $t_{\text{handling,load}} = 1$ h and $t_{\text{handling,unload}} = 0.5$ h are assumed so that the roundtrip time $t_{\text{roundtrip}}$ is 84.9 h.

The onshore intermediate storage must be capable of storing the CO_2 stream coming from the liquefaction plant, $\dot{m}_{\text{source}}$ while no ship is moored for loading. The capacity of the onshore intermediate storage depends on $\dot{m}_{\text{source}}$ and the time period between the end of the loading process for the current ship and the arrival of the next ship. This time period is equal to the difference between the roundtrip time divided by the number of ships (n_{ship}) and the loading time ($t_{\text{load,ship}}$). Thus, the required net onshore intermediate storage capacity is equal to:

$$M_{\text{stor,net}} = \dot{m}_{\text{source}} \cdot \left(\frac{t_{\text{roundtrip}}}{n_{\text{ship}}} - t_{\text{load,ship}} \right) \tag{3.12}$$

For the exemplary values shown in Figure 27, the resultant net onshore intermediate storage capacity is $M_{\text{stor,net}} = 4197$ t. The corresponding gross onshore intermediate storage capacity can be calculated from $M_{\text{stor,net}}$ and the CO_2 densities for the filled and empty storage tanks:

$$M_{\text{stor,gross}} = \frac{M_{\text{stor,net}}}{1 - \rho_{\text{CO2,empty}}/\rho_{\text{CO2,filled}}} \tag{3.13}$$

For the exemplary values shown in Figure 27, liquid CO_2 with a density of $\rho_{\text{CO2,filled}}$ (1.15 t/m^3 at -50 °C and 6.8 bar) is assumed for the filled storage tank and gaseous CO_2 with a density of $\rho_{\text{CO2,empty}}$ (0.018 t/m^3 at -50 °C and 6.8 bar) for the empty tank. Thus, the gross onshore intermediate storage capacity $M_{\text{stor,gross}}$ is 4262 t.

It is assumed that the unloading process occurs at constant pressure so that the empty tank pressure is still equal to the transport pressure. Generally, though, the model does not specify how the unloading process is carried out as $\rho_{\text{CO2,empty}}$ can be freely selected. In the same manner, the full tank density $\rho_{\text{CO2,filled}}$ can be chosen to be less than the actual liquid CO_2 density if the tank will not be completely filled.

When the ship is moored at the harbour, the ship tanks are filled with a mass flow rate of $\dot{m}_{\text{load,ship}}$ from the onshore intermediate storage. It has been taken into account that within this time period, the onshore intermediate storage is still being filled with a mass flow rate of $\dot{m}_{\text{source}}$ from the liquefaction plant. The loading process of the ship and the onshore intermediate storage unloading process are calculated separately. The unloading time of the onshore intermediate storage is:

$$t_{unload,stor} = \frac{M_{stor,net}}{\dot{m}_{load,ship} - \dot{m}_{source}} \tag{3.14}$$

The loading time of the ship is determined from net ship capacity $M_{ship,net}$ and the ship loading mass flow rate $\dot{m}_{load,ship}$:

$$t_{load,ship} = \frac{M_{ship,net}}{\dot{m}_{load,ship}} \tag{3.15}$$

For a feasible transport chain, the unloading time of the onshore intermediate storage and the loading time of the ship must be equal:

$$t_{unload,stor} = t_{load,ship} \tag{3.16}$$

The exemplary values shown in Figure 27 are selected so that the loading time of the ship (and thus, the unloading time of the onshore intermediate storage) is 0.5 h. This results in a net ship capacity $M_{ship,net}$ of 4248 t and a ship loading mass flow rate $\dot{m}_{load,ship}$ of 8496 t/h. $\dot{m}_{load,ship}$ is determined by the loading mass flow rate per tank and the number of tanks:

$$\dot{m}_{load,ship} = \dot{m}_{load,tank} \cdot n_{tank} \tag{3.17}$$

Since the number of tanks is 2 in Figure 27, the loading mass flow rate per ship is twice as high as the loading mass flow rate per tank (4248 t/h). Similarly, $M_{ship,net}$ is determined by the number of tanks n_{tank} and their net capacity $M_{tank,net}$ (2124 t):

$$M_{ship,net} = M_{tank,net} \cdot n_{tank} \tag{3.18}$$

The gross ship capacity $M_{ship,gross}$ can be calculated from $M_{ship,net}$ in equivalence to eq. (3.13). For the dimensioning of the transport chain components, it is not relevant whether $M_{ship,net}$ and $M_{ship,gross}$ consist of a small number of tanks and a large tank capacity or vice versa. However, the number of tanks and their individual capacity are relevant for calculating the boil-off gas quantity during transport as described in section 3.3.1.

The total quantity of CO_2 transported per year is

$$\dot{M}_{trans} = \frac{8760\,\text{h/a}}{t_{roundtrip}} M_{ship,net} \cdot n_{ship} \tag{3.19}$$

The quantity of CO_2 transported $\dot{M}_{trans}$ must be equal to the mass flow rate coming from the pipeline, $\dot{m}_{source}$:

$$\dot{m}_{source} = \dot{M}_{trans} \tag{3.20}$$

For the exemplary values shown in Figure 27, $\dot{M}_{trans}$ is 0.876 Mt/a, which corresponds with the source mass flow rate of 100 t/h.

As illustrated in Figure 27, two strategies are possible for unloading and injection: The CO_2 can either be directly injected into the geological CO_2 storage, so that unloading and injection represent the same process, or it can be unloaded into offshore intermediate storage so that the ship can return to the harbour while injection is still continuing (q. v. section 2.5). The benefit of using offshore intermediate storage is that the time required for unloading the ship is not limited by the injection mass flow rate since both processes are separated. Independent of the unloading and injection strategy, the mass flow rates are defined by the mass flow rate per tank or per well, respectively, and the number of tanks and wells:

$$\dot{m}_{unload,ship} = \dot{m}_{unload,tank} \cdot n_{tank} \tag{3.21}$$

$$\dot{m}_{injection} = \dot{m}_{injection,well} \cdot n_{well} \tag{3.22}$$

The time required for unloading and the time required for injection are defined as follows:

$$t_{unload,ship} = \frac{M_{ship,net}}{\dot{m}_{unload,ship}} \tag{3.23}$$

$$t_{injection} = \frac{M_{ship,net}}{\dot{m}_{injection}} \tag{3.24}$$

For direct injection, $\dot{m}_{unload,ship}$ is equal to $\dot{m}_{injection}$ and $t_{unload,ship}$ is equal to $t_{injection}$. This is the case for the example shown in Figure 27, where $\dot{m}_{unload,ship} = \dot{m}_{injection} = 450$ t/h and $t_{unload,ship} = t_{injection} = 9.4$ h.

If offshore intermediate storage is used, its net capacity is determined by the unloading time $t_{unload,ship}$ and difference between the unloading and injection mass flow rates. Thus, it is assumed that injection of CO_2 is also carried out during unloading:

$$M_{ofstor,net} = t_{unload,ship} \cdot \left(\dot{m}_{unload,ship} - \dot{m}_{injection} \right) \qquad (3.25)$$

Usually, multiple ships are used for transport and it is assumed that these ships operate on the same schedule with equal distance to each other, have the same capacity and the same loading, unloading and injection mass flow rates. At any given time, only one single ship may be moored at the harbour or injection site. The time period between the mooring of one ship and the next is equal to the roundtrip time divided by the number of ships. The transport chain is only viable if the loading time, unloading time and injection time are shorter than the time period between the arrival of the current and the following ship:

$$t_{load,ship} < \frac{t_{roundtrip}}{n_{ship}} \qquad (3.26)$$

$$t_{unload,ship} \leq t_{injection} < \frac{t_{roundtrip}}{n_{ship}} \qquad (3.27)$$

57

4 Results

4.1 CO$_2$ Liquefaction Process

4.1.1 Refrigerant Selection

An evaluation of different refrigerants for the closed cycle liquefaction process has been conducted. For this purpose, the minimum specific energy demands of the 1-stage, the 2-stage and the 3-stage closed cycle process have been determined for various common refrigerants. The term 'minimum' indicates that the simplified processes described in section 3.2.1 have been used, in which, among other things, the energy demand of the seawater cooling pump is neglected. Secondly, it means that the intermediate CO$_2$ pressure p$_1$ (in the case of the 2-stage process) and the intermediate CO$_2$ pressures p$_1$ and p$_2$ (in the case of the 3-stage process) have been optimised with respect to the total specific energy demand of the liquefaction process. A sensitivity analysis has been carried out to determine the optimum values for p$_1$ (2-stage process) or p$_1$ and p$_2$ (3-stage process). The optimum intermediate CO$_2$ pressure depends on the process design (2-stage or 3-stage, with or without cascade heat exchanger and refrigerant type) and boundary conditions such as the CO$_2$ input and output pressure and the ambient temperature. Besides the boundary conditions, the intermediate CO$_2$ pressure is the only degree of freedom for the multi-stage processes. The intermediate CO$_2$ pressure is discussed in more detail in section 4.1.3.

In Figure 28, a sensitivity analysis for the 3-stage NH$_3$-NH$_3$-Propene process is shown as an example. The notation NH$_3$-NH$_3$-Propene means that ammonia has been used for the two upper temperature cycles while Propene (Propylene) has been used for the lower temperature cycle. For the liquefaction of pure CO$_2$ at -50 °C, the optimum intermediate pressures are $p_2 = 14.5$ bar and $p_1 = 28$ bar and the minimum specific energy demand is 12.4 kWh/t CO_2.

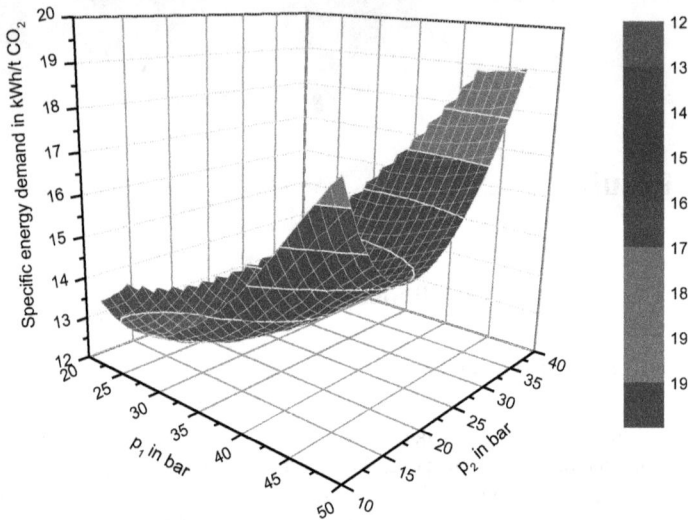

Figure 28: Sensitivity analysis for the specific energy demand of the 3-stage NH_3-NH_3-Propene closed cycle process for the liquefaction of pure CO_2 at -50 °C

Figure 29 shows the minimum specific energy demands of the 1-stage, the 2-stage and the 3-stage closed cycle process for the liquefaction of pure CO_2 at -50 °C using various common refrigerants. The minimum specific energy demands of the 1-stage and the 3-stage open cycle process are shown for comparison. The results show that the energy demand mainly depends on the process design (e.g. 1-stage, 2-stage or 3-stage) and - in the case of the 2-stage and the 3-stage closed cycle processes - only to a small extent on the refrigerant used. For the 1-stage closed cycle process, values between 22.1 kWh/t CO_2 and 33.7 kWh/t CO_2 have been calculated. The minimum specific energy demand for the 2-stage closed cycle process is generally approximately 40 % lower than for the 1-stage closed cycle process with values between 13.5 kWh/t CO_2 and 14.5 kWh/t CO_2. Values between 12.5 kWh/t CO_2 and 12.4 kWh/t CO_2 have been determined for the 3-stage closed cycle process. The efficiency gain for an additional third cycle is much lower in comparison with an energy demand reduction of approximately 10 %. A similar, but generally higher energy demand was found both for the 1-stage and the 3-stage open cycle process. Moreover, the liquefaction of CO_2 with impurities would require significant additional effort when using an open cycle process (q.v. section 3.2.2). Therefore, further studies of the open cycle process have not been conducted.

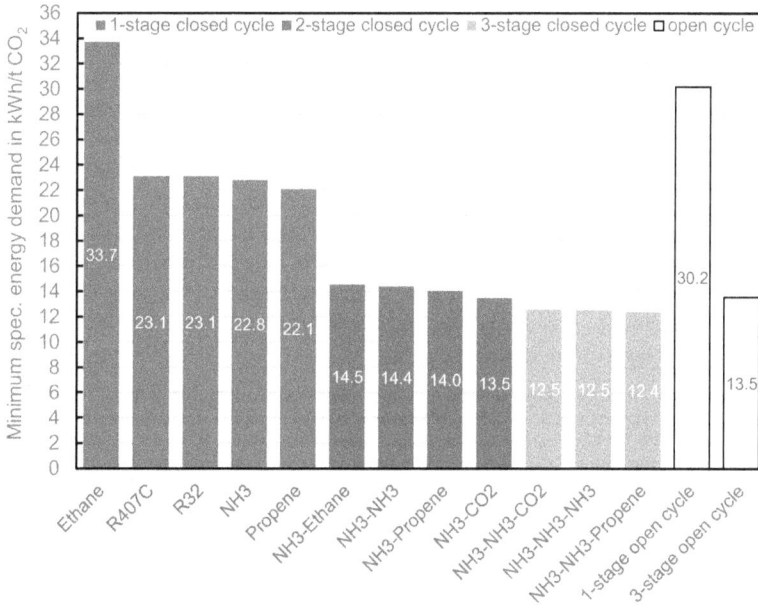

Figure 29: Minimum specific energy demands of the 1-stage, the 2-stage and the 3-stage closed cycle process using various common refrigerants for liquefaction at -50 °C. The minimum specific energy demands of the 1-stage and the 3-stage open cycle process are shown for comparison.

Besides the energy demand, other factors such as environmental aspects, health and safety and operational issues are usually taken into account for refrigerant selection. Common criteria for selection have been presented in section 2.1.1. These criteria have been quantified using six parameters. Two key aspects for environmental impact analysis of refrigerants are the Global Warming Potential (GWP) and the Ozone Depletion Potential (ODP). The GWP quantifies the impact of a refrigerant on the greenhouse effect in relation to the greenhouse effect of CO_2, usually in a time frame of 100 years. The ODP quantifies the contribution of the refrigerant towards

61

the depletion of the natural ozone layer in relation to the reference refrigerant R11 (Trichlorofluoromethane). All refrigerants considered in this work have an ODP of zero, thus the GWP is used to assess the environmental impact. A key parameter for the safety of a refrigerant is the low flammability limit (LFL), which is the minimum refrigerant concentration at which a refrigerant-air-mixture can be ignited. An important parameter for health impact is the time-weighted threshold limit value (TLV) to which workers can be exposed in their everyday jobs. To evaluate a refrigerant with respect to operational issues, the triple, boiling and critical points are significant: The triple point must be outside of the working range of the refrigerant to avoid the solid phase formation. The boiling temperature at atmosphere pressure should be as low as possible, preferably lower than -50 °C for CO_2 liquefaction to avoid a vacuum in the refrigerant cycle. The critical temperature must be higher than the cooling medium (e.g. seawater) temperature because phase change – and thus condensation of the refrigerant - only occurs between the triple and the critical temperature. For a given critical temperature, the critical pressure should be as low as possible to reduce the investment costs of the components such as the compressor, condenser and piping.

Table 8: Six key parameters used to select the refrigerant for CO_2 liquefaction. The abbreviations are: GDP - Global warming potential; LFL - Low flammability limit; TVL - Time weighted average threshhold value; (*) - CO_2 Sublimation temperature: Pressure at triple point is 5.1 bar, thus a phase change between solid and gaseous occurs at 1 bar. Hence, no boiling temperature at 1 bar is defined.

	Environment	Health and Safety		Physical properties		
	GDP	LFL in vol.-%	TLV in ppm	Triple pt. temp. in °C	Boiling temp. at 1 bar in °C	Critical pt. in °C / bar
R32	675	14.5	1000	-137	-52	78/ 58
R407C	1774	-	1000	-73	-44	86 / 46
Ethane	6	2.9	1000	-182	-88	32 / 49
Propene	2	2.4		-185	-48	91 / 46
CO_2	1	0	5000	-57	-78 (*)	31 / 78
NH_3	0	15	25	-77	-33	132 / 13

In Table 8, these six key parameters are shown for the considered refrigerants. R32 (Difluoromethane) and R407C (a mixture of different hydrofluorocarbons) are widely used in industry. They are characterized by their high GWP of 675 and 1774, respectively. Since the purpose of a CCS project is to mitigate the greenhouse effect, these refrigerants are considered to be unsuitable for the liquefaction of CO_2 and, thus, were not studied further. Moreover, the results for the 1-stage closed cycle process in Figure 29 show that other refrigerants such as ethane, propene and ammonia lead to an equally low energy demand. CO_2 was not considered as a refrigerant for the 1-stage process since the trans-critical refrigeration cycle requires high pressure components and thus, is usually only employed when health aspects or lack of space are arguments in its favour (such as in the food or car industries). For the 2-stage closed cycle process, only small differences between ethane, propene, CO_2 and ammonia could be found when using these refrigerants for the lower temperature cycle. In all four cases, ammonia was used for the upper temperature cycle. Ethane has the lowest boiling temperature among the four considered lower temperature cycle refrigerants. The critical temperature is just high enough for seawater cooling in the temperate climate zone but is too low for air cooling in general or seawater cooling in other climates. Moreover, with a GWP of 6, its contribution to the greenhouse effect is significant. For this reason, only propene, CO_2 and ammonia are compared for the 3-stage closed cycle process.

Figure 29 shows that a similar specific minimum energy demand can be obtained with these three refrigerants. The triple point temperature of CO_2 (-57 °C) is close to the liquefaction temperature for the CO_2 stream. When taking into account that a certain minimum internal temperature approach in the heat exchanger (3 K – 5 K) is necessary between the refrigeration cycle and the CO_2 product stream at -50 °C, solid CO_2 (dry ice) could form in the refrigeration cycle when small pressure changes occur and CO_2 is used as the refrigerant. The lowest energy demand has been found for propene, but the difference with ammonia is minor (< 1 %). However, propene has a GWP of 2, a very low LFL of 2.4 vol.-% and a lower critical temperature as well as a higher critical pressure than ammonia. With the exception of the lower boiling temperature and the higher TLV, the properties of propene are less suitable for the liquefaction of CO_2 than the properties of ammonia. Ammonia is toxic so a rather low TLV is tolerable, but it is detectable by its odour at even small concentrations (5 ppmv) while being toxic only at much higher concentrations. Since ammonia is widely used in industry, well established safety regulations can be followed.

In comparison to ammonia, the energetic advantage of using propene for the lower temperature cycle is small. There is even an energetic disadvantage when taking into account that the minimum specific energy demand of a multi-stage closed cycle process can be reduced when only one single refrigerant is used for all cycles (q. v. section 4.1.3). Hence, for all further studies in this work, ammonia is used as the refrigerant for all cycles.

4.1.2 Impact of CO₂ Impurities

The impact of CO_2 impurities on the minimum specific energy demands of the liquefaction processes is studied for the five CO_2 streams introduced in Chapter 3: A pure CO_2 stream, a CO_2 stream from a Post-Combustion CO_2 capture plant (Post), two CO_2 streams from Oxyfuel CO_2 capture plants with different CO_2 purities (Oxy98 and Oxy96) and a CO_2 stream based on a cluster of different CO_2 emitters (CL-MIN-CO2). The minimum specific energy demands of the 1-stage, the 2-stage and the 3-stage closed cycle process for the liquefaction of those streams are shown in Figure 30. While results for the liquefaction of pure CO_2 have already been presented in section 4.1.1, the results found in Figure 30 are generally 0.2 kWh/t CO_2 higher. This can be attributed to the fact that superheaters have been added to ensure 5 K superheat at the ammonia compressor inlets (heat exchangers *PH1* and *PH2* in the 2-stage closed cycle base process shown in Figure 15, and *PH1*, *PH2* and *PH3* in the 3-stage closed cycle base process shown in Figure 17). While this reduces the efficiency of the process, superheating is usually done to avoid liquid droplets at the compressor inlet. For all results shown in this section and in the following, superheating at the compressor inlet is assumed. The results without superheat can be found in another publication [74].

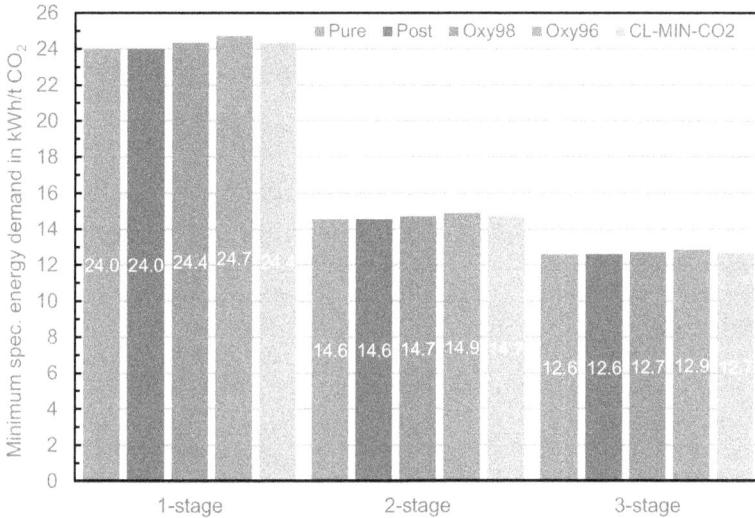

Figure 30: Minimum specific energy demands of the 1-stage, the 2-stage and the 3-stage closed cycle process for different CO₂ streams with typical impurity concentrations

Figure 30 shows that the greatest differences in the minimum specific energy demand can be attributed to the number of refrigeration cycle stages (1-stage, 2-stage or 3-stage) rather than the composition of the CO_2 streams: With a 2-stage process, the energy demand can be reduced by approximately 39 % with respect to the 1-stage process. The 3-stage process leads to a further reduction by 13 % concerning the 2-stage process. The energy demand slightly increases with higher impurity concentrations. Thus, the lowest minimum specific energy demand was found for the pure and the Post CO_2 stream and the highest was found for the Oxy96 CO_2 stream. The increase in the energy demand decreases with a higher number of stages: The highest additional energy demand for each CO_2 stream was found for the 1-stage and the lowest was found for the 3-stage process. However, the differences between the 2-stage and the 3-stage process are marginal. For the Oxy96 CO_2 stream, the additional minimum specific energy demand with respect to pure CO_2 is approx. 0.7 kWh/t CO_2 for the 1-stage process, 0.33 kWh/t CO_2 for the 2-stage process and 0.27 kWh/t CO_2 for the 3-stage process. This corresponds to an increase in the minimum specific energy demand of approximately 3 % for the 1-stage and approx. 2 % for the 2-stage and the 3-stage process with respect to pure CO_2. For all

65

other CO_2 streams, the increase is lower due to the lower impurity concentrations and remains below 1 %. Thus, it can be concluded that the additional minimum specific energy demand due to the impurities is almost negligible for the considered CO_2 streams.

In Figure 29 and Figure 30, the minimum specific energy demand of the 1-stage closed cycle process is merely shown for comparison since it would not be used in reality. This is not only due to its high energy demand, but also due to the high pressure ratio of the ammonia compressor and the resulting high discharge temperature. Twin-rotary screw compressors are usually used for large-scale ammonia systems, which have a maximum single-stage pressure ratio of 20 [35]. The discharge temperature may not be much higher than 100 °C as typical lubrication oils carbonise at approx. 150 °C. In Table 9, the pressure ratios, the specific electrical energy demand of the ammonia compressor and the discharge temperatures are shown for the pure and the Oxy96 CO_2 stream. The results for Post, Oxy98 and CL-MIN-CO_2 are higher than for pure CO_2 but lower than for the Oxy96 CO_2 stream and can be found in appendix A.1.

Table 9: Pressure ratio, specific electrical energy demand and compressor discharge temperature for the liquefaction of a pure and an Oxy96 CO_2 stream using the 1-stage, the 2-stage and the 3-stage closed cycle process. The indices 'c1', 'c2' and 'c3' refer to the refrigeration cycles, counting from upper to lower temperature. The compressor discharge temperature in cycle II is T_{IIe} in the case of the 2-stage closed cycle process (Figure 15) and T_{IIf} in the case of the 3-stage closed cycle process (Figure 17).

		Pressure ratio of compressors			Spec. el. energy demand of comp. in kWh/t CO_2			Comp. discharge temperature in °C		
		π_{CI}	π_{CII}	π_{CIII}	w_{CI}	w_{CII}	w_{CIII}	T_{Ia}	$\frac{T_{IIe}}{T_{IIf}}$	T_{IIIe}
1-stage	Pure	25.1			24.0			257		
	Oxy96	25.1			24.7			257		
2-stage	Pure	5.0	5.8		10.6	4.0		123	93	
	Oxy96	4.9	5.9		10.7	4.1		122	95	
3-stage	Pure	3.1	2.9	3.6	7.4	3.5	1.7	92	58	50
	Oxy96	3.0	2.9	3.8	7.4	3.6	1.9	90	58	54

As expected, the values in Table 9 mainly depend on the process design (1-stage, 2-stage or 3-stage) and only to a small extent on the quality of the CO_2 stream. With discharge temperatures of up to 257 °C, the maximum value of 150 °C is exceeded for the 1-stage process. Likewise, the pressure ratio of 25 is higher than the usual maximum value for single-stage screw compressors. The pressure ratios of the 2-stage and the 3-stage process are much lower with values of 4.9 to 5.9 and 3.0 to

3.8, respectively. The compressor discharge temperatures of the 2-stage process are still higher than 100 °C, but below 150 °C. In reality, the discharge temperatures would be lower as the cooling effect of the lubrication oil is not considered in the model. For this reason, both the 2-stage and the 3-stage process are considered to be viable options for CO_2 liquefaction.

In chapter 2, a rule by Wang [35] was introduced which suggests that the same pressure ratio is usually used for each stage of a multi-stage refrigeration process (eq. (2.2)). According to the results shown in Table 9, this is not the optimum configuration for the analysed closed cycle liquefaction process. The results suggest that using the same pressure ratio for each stage can merely be considered as a rule of thumb rather than a universally applicable optimum configuration.

In equivalence to the results in Table 9, no major differences between the CO_2 streams were found in regard to other results of the refrigeration cycle side of the process. This can be explained by the fact that both the heat quantity transferred from the CO_2 stream to the refrigeration cycle (i.e. the cooling demand required for liquefaction) and the temperature levels of the refrigeration cycle stages are approximately the same for all considered CO_2 streams. The cooling demand is determined by the heat of evaporation of the individual CO_2 stream components and their concentration in the CO_2 stream. Since all considered CO_2 streams mainly consist of CO_2, only slight differences in the cooling demand are observable.

One notable difference between pure CO_2 and the CO_2 streams with impurities is the temperature gradient during condensation and evaporation. While the temperature of pure CO_2 remains constant during condensation, the temperature of a CO_2 stream with impurities is continuously declining. This can be attributed to the fact that in the condensation of a multi-component mixture, all components condense at a given pressure, but those with a higher boiling temperature condense at a higher rate than those with a lower boiling temperature. Thus, the concentration of those components with a high boiling temperature decreases in the remaining gas phase while the concentration of the components with a low boiling temperature increases. Consequently, the temperature of the vapour-liquid equilibrium is continuously declined during condensation until the bubble temperature of the mixture is reached. These relationships can be represented in a Q, T diagram.

The Q, T diagram of the CO_2 stream side is shown in Figure 31 for the 2-stage closed cycle process. In this diagram, the relationships described above can be observed: The cooling demand (i.e. the specific heat quantity) for liquefaction slightly increases with an increasing impurity concentration. With the exception of pure CO_2, a temperature gradient during condensation - also referred to as condensation curve – can be observed for all CO_2 streams. Three temperature drops occur: The first one at $Q = 0$ MJ/t corresponds with the expansion from input pressure to p_{liq} (as defined in Figure 15 and Figure 17). This expansion is merely done to allow the comparison of the closed cycle base processes with the optimised versions discussed in section 4.1.3. The temperature drop of approx. 5 K can be explained by the positive Joule-Thompson coefficient for liquid CO_2 at these temperature and pressure conditions [91].

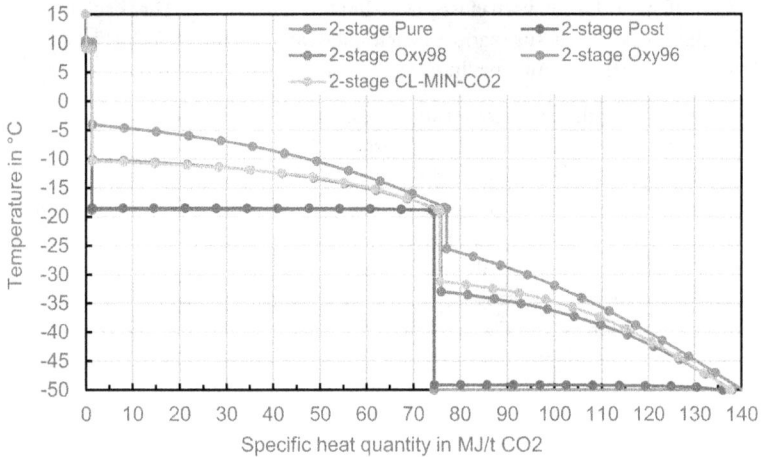

Figure 31: Q, T - diagram of the 2-stage closed cycle process for various CO_2 streams. Only the CO_2 stream side is shown.

The second temperature drop at $Q \approx 0$ MJ/t corresponds with the expansion from p_{liq} into two-phase region at p_1. The CO_2 stream is separated into a liquid and a vaporous phase. This is an isenthalpic process and is also called flash evaporation. A so called vapour-liquid equilibrium is reached, in which the volatile components such as H_2, O_2, and N_2 have a higher concentration in the vapour phase than in the liquid phase. The vapour fraction x after expansion depends on the enthalpies of the liquid and the vaporous phase at p_1. As the enthalpy after expansion is equal to the enthalpy before expansion, x can be calculated when the enthalpies of the liquid and the vaporous phase are known. The temperature in vapour-liquid equilibrium is generally higher than the bubble temperature (no vapour present) and lower than the dew point temperature of the mixture (only vapour present). The temperature can be directly determined if p_1 and the enthalpy of the mixture are known. Calculation of the liquid and vapour phase concentrations, the vapour fraction and temperature is done by iteratively solving the Rachford-Rice equation [92], which is a formulation of the vapour-liquid equilibrium of a multi-component mixture. The second temperature drop is therefore not only caused by the Joule-Thompson effect, but mainly due to the fact that the CO_2 is expanded into a two-phase regime.

The same explanation applies for the expansion from p_1 to p_{out} which corresponds with the third temperature drop at $Q \approx 75\,MJ/t$ CO_2. The second and the third temperature drop are both approximately 30 K for pure CO_2. For CO_2 with impurities, the temperature drops decrease with increased impurity concentration. This can be explained by the fact that the dew point temperature of the CO_2 stream with a high fraction of impurities is higher. Moreover, the temperature drops with impurities are generally higher for the second expansion from p_{liq} to p_1 than for the third from p_1 to p_2. This is a consequence of the fact that the enthalpy of evaporation (which is the difference between the liquid and vaporous phase enthalpies) increases for lower pressures. For a higher enthalpy of evaporation, less vapour is formed during expansion and a lower temperature drop occurs.

The Q, T diagram of the CO_2 stream side process is shown in Figure 32 for the 3-stage closed cycle. While the cooling demand is the same as in the 2-stage closed cycle process, a different temperature profile can be observed. Most notably, four instead of three temperature drops occur, which can be explained by the additional refrigeration cycle and the associated additional expansion step. The temperature drops show the same general relationships as in the 2-stage closed cycle process. For pure CO_2, each temperature drop is approximately 20 K. For CO_2 with impurities, the temperature drops decrease with increased impurity concentration and are highest for the expansion from p_{liq} to p_1 at Q = 0 MJ/t and lowest for the expansion from p_2 to p_3 at Q $\approx$ 96 MJ/t.

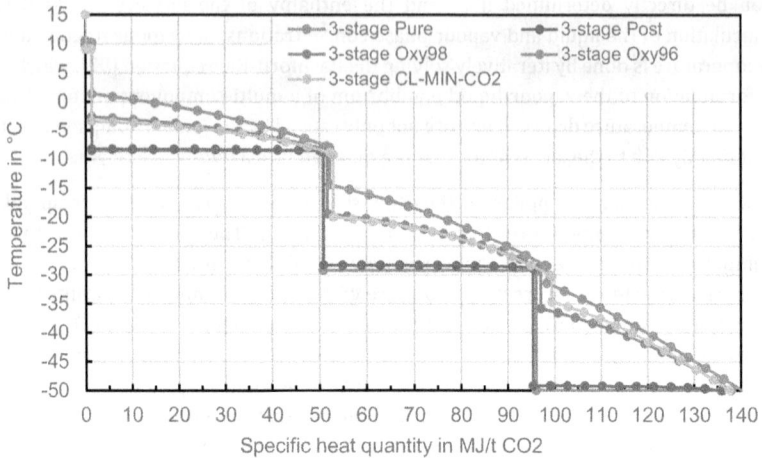

Figure 32: Q, T - diagram of the 3-stage closed cycle process for various CO_2 streams. Only the CO_2 stream side is shown.

The results shown in Figure 31 and Figure 32 have several implications for the design of the CO_2 stream/refrigerant heat exchangers (CO_2 condensers). On the one hand, the cooling demand increases with an increased impurity concentration, on the other hand, the mean temperature gradient between the CO_2 stream and the refrigerant increases. The mean temperature gradient over the heat exchanger depends on the CO_2 stream and the refrigerant inlet and outlet temperatures. While the CO_2 inlet temperature increases with an increased impurity concentration, the CO_2 outlet temperature remains approximately the same. Since the refrigerant inlet temperature (cold side inlet) depends on the CO_2 outlet temperature (hot side outlet), it also remains approximately the same. At the same time, the refrigerant outlet temperature is the same as the refrigerant inlet temperature since evaporative cooling is used. Thus, only the CO_2 inlet temperature increases which results in an increase in the mean temperature difference. The consequences can be illustrated by considering the logarithmic mean temperature difference (LMTD) equation $Q = UA \cdot \Delta T_m$. The LMTD (ΔT_m) depends on the flow configuration of the heat exchanger (e.g. parallel or counter flow). The higher the value for ΔT_m, the lower the UA-value necessary to transfer a certain quantity of heat. In Table 10, these parameters are listed for the liquefaction processes and CO_2 streams shown in Figure 31 and Figure 32. The UA-value is shown as a relative quantity with respect to the UA-value for pure CO_2. It can be seen that the relative UA-values in Table 10 significantly decrease with an increased impurity concentration. On the other hand, the CO_2 pressures p_1, p_2 and p_3 must be higher for higher impurity concentrations to allow the condensation of all CO_2 stream components. For the Oxy96 CO_2 stream, for example, the UA-value of the heat exchanger $HE2$ (as defined in Figure 15 and Figure 17) is only 11 % of the UA-value for pure CO_2. At the same time, the output pressure p_3 must be raised from 6.77 bar in the case of pure CO_2 to 23.7 bar for the Oxy96 CO_2 stream. The increase in ΔT_m and the associated decrease in the UA-value is slightly offset by the increase in the cooling demand Q (which leads to a higher UA-value), but in proportion the increase of Q is very small. Thus, a lower UA-value is reached in total when impurities are present. However, a decrease of the UA-value does not necessarily imply that the heat transfer surface area A can be decreased: Generally, the presence of impurities in the vaporous phase leads to a deterioration in the convective heat transfer and, as a result, to a decrease of the overall heat transfer coefficient U. A more detailed study of the local convective heat transfer coefficients during condensation of CO_2-rich mixtures would be necessary to estimate the impact of impurities on the overall heat transfer coefficient U, which is not within the scope of this work. This would also include a calculation of the local temperature differences over the length of the heat exchanger, which provides more accurate results for condensers than the utilisation of the logarithmic mean temperature difference ΔT_m [93]. For the accurate design of mixed-vapour condensers, different methods such as the Silver, Silver-Bell and Colburn methods are available in literature [94].

71

Table 10: Main design parameters of the CO_2/refrigeration cycle heat exchangers in the 2-stage and the 3-stage closed cycle process for various CO_2 streams: Intermediate and output pressure of CO_2, the logarithmic mean temperature difference (LMTD), heat duty and the UA-values of the heat exchangers *HE1*, *HE2* and *HE3* relative to the UA-values for pure CO_2. For the LMTD, counter-current flow is assumed. p1,p2, p3, Q1, Q2, Q3 and *HE1*, *HE2* and *HE3* are defined in Figure 7 and Figure 8.

No. of stages, CO₂ stream	CO₂ pressure in bar			LMTD			Cooling demand in MJ/t CO₂			UA relative to the value for pure CO₂ in %		
	p_1	p_2	p_3	$\Delta T_{m,1}$	$\Delta T_{m,2}$	$\Delta T_{m,3}$	Q_1	Q_2	Q_3	$\frac{UA_1}{UA_1^p}$	$\frac{UA_2}{UA_2^p}$	$\frac{UA_3}{UA_3^p}$
2 Pure	20.25	6.77		3.2	2.9		73	17		100	100	100
Post	20.50	7.03		3.4	3.8		73	17		95	77	
Oxy98	28.75	15.03		6.6	19.9		74	17		49	15	
Oxy96	37.00	23.71		8.5	27.4		75	18		39	11	
CL-MIN-CO₂	27.75	15.40		6.5	21.7		74	17		50	14	
3 Pure	27.50	14.50	6.78	3.0	3.0	3.0	49	45	27	100	100	100
Post	27.75	15.00	7.03	3.1	3.4	3.8	49	44	27	100	86	100
Oxy98	36.00	23.25	15.03	5.3	12.3	17.2	50	45	27	60	24	22
Oxy96	44.00	31.75	23.71	6.5	17.0	21.5	50	46	28	49	18	18
CL-MIN-CO₂	34.25	22.00	15.40	5.3	13.1	18.3	51	46	27	62	23	21

A strategy to dampen the reduction in U due to impurities would be not to fully condense the stream and to vent a certain small fraction of the vaporous phase, which contains a higher concentration of impurities than the liquid phase. Venting of the vaporous phase would lead to a higher CO_2 concentration in the liquid stream after condensation but is also associated with a lower quantity of CO_2 being transported and stored. This has not been studied in this work since it deals with the transport of CO_2 streams with different impurity concentrations and purification is already carried out at the gas processing unit of the Oxyfuel plant.

Although the pressure of the CO_2 stream increases for increased impurity concentrations (Table 10), the CO_2 outlet temperatures are approximately the same for all CO_2 streams. Due to this fact, the refrigerant temperatures and pressures remain the same, regardless of the CO_2 stream quality. It can be concluded that the CO_2 stream composition has a significant impact on the heat exchanger design, which can be attributed to the differences of the CO_2 pressure and heat exchanger inlet temperatures. At the same time, the pressure and temperature of the refrigerant do not change significantly.

4.1.3 Optimisation Measures

Five different measures of improvements have been investigated for the closed cycle liquefaction process. Two measures concern the CO_2 stream side, and three measures the refrigeration cycle side. For the CO_2 stream side, energy recovery via a liquid phase and a two-phase CO_2 expander are analysed. For the refrigeration side, the replacement of the cascade heat exchanger by a phase separator, the use of a compressor aftercooler and the replacement of the refrigerant superheater by an internal heat exchanger are studied. The working principles of these five measures are described in more detail in section 3.2.1.

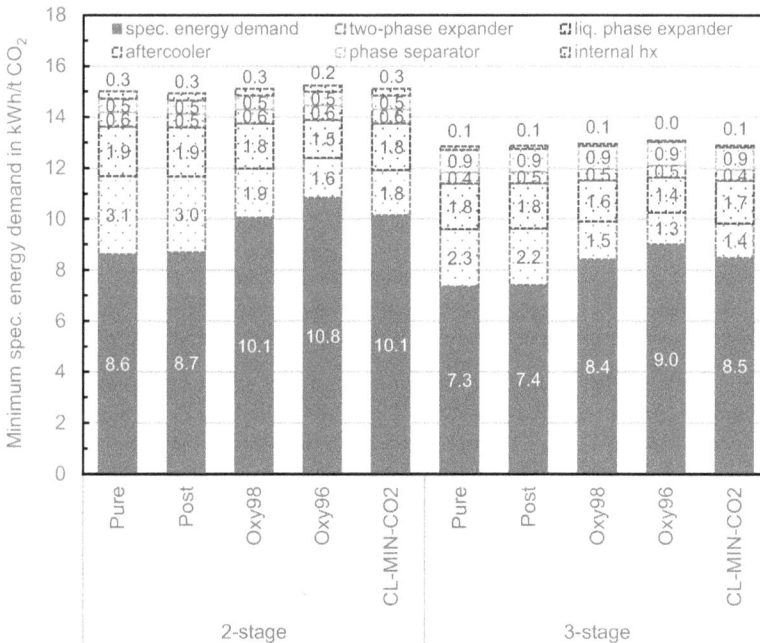

Figure 33: Minimum specific energy demands of the optimised 2-stage and the optimised 3-stage closed-cycle process for various CO_2 streams

In Figure 33, the impact of these measures on the minimum specific energy demands is shown. The values in the blue bars represent the minimum specific energy demands of the optimised processes with all five measures implemented, while the

73

values in the dotted bars show the reduction in the energy demand obtained from one individual optimisation measure when only this particular measure is implemented. It can be seen that the minimum specific energy demands of the optimised processes (values in the blue bars) are approximately 27 % (Oxy96) to 41 % (pure CO_2) lower than the minimum specific energy demands of the respective base processes (shown in Figure 14).

However, the combination of these five measures is not as effective as the sum of the individual measures: The difference between the specific energy demand of the base processes and the optimised versions with all five measures implemented is lower than the sum of the individual energy reductions obtained from each measure (values in the dotted bars). For example, in the case of the 3-stage process with pure CO_2, the sum of the individual optimisation measures (values in the dotted bars) is 5.5 kWh/t CO_2, but the difference between the minimum specific energy for the base process and the optimised process is 5.3 kWh/t CO_2 (12.6 kWh/t CO_2 – 7.3 kWh/t CO_2). This means that the combination of all five measures slightly reduces the impact of each individual measure. For example, the aftercooler has a greater effect on the base process as the pressure ratio of the ammonia compressors is higher and consequently, the discharge temperature is higher than in the optimised process. The pressure ratios in the optimised process are lower due to the use of the phase separator, which removes the need for differences between the lower temperature refrigeration cycle condensation pressures and temperatures and the upper temperature cycle evaporation pressures and temperatures.

While there was hardly any impact of the CO_2 stream compositions found for the minimum specific energy demands of the closed cycle base processes, a greater influence can be seen for the optimised processes. More precisely, a significant impact of the CO_2 stream composition can be observed in Figure 33 for the liquid and the two-phase expanders, which concern the CO_2 stream side. With an increased impurity concentration, less energy can be recovered by the expanders, especially by the two phase expanders. This effect can be explained by the increasing bubble pressure of CO_2 which leads to lower pressure ratios over the expanders. For example, the total pressure ratio of the input pressure (p_{in}) to the output pressure is $\pi = \frac{100\ bar}{6.77\ bar} = 14.8$ in the case of pure CO_2, but only $\pi = \frac{100\ bar}{23.71\ bar} = 4.2$ for the Oxy96 CO_2 stream. The total pressure ratio can be divided into the pressure ratio of the liquid expander, π_{liq}, and the pressure ratios of the two-phase expanders, π_1 and π_2 for the 2-stage closed cycle process and π_1, π_2 and π_3 for the 3-stage closed cycle process. The pressure ratio of the liquid expander only depends on the CO_2 stream and not on the process design. With values between 2.2 for pure CO_2 and

1.7 for the Oxy96 CO_2 stream, the variation is comparably low so that similar reductions in the energy demand are obtained. As π_{liq} does not depend on the process design, it could be expected that the energy demand reduction is the same for the 2-stage and the 3-stage cycle. According to the results shown in Figure 33, this is not the case. This means that other factors besides the pressure ratio have an impact on the reduction in the energy demand due to the liquid expander, which is discussed below. The pressure ratios of the 2-stage closed cycle process' two-phase expanders are $\pi_1 = 2.0$ and $\pi_2 = 3.4$ in the case of pure CO_2 and $\pi_1 = \pi_2 = 1.6$ for the Oxy96 CO_2 stream. This shows a much greater dependency of the CO_2 stream composition for the pressure ratios of the 2-stage expanders. Consequently, the energy reduction obtained from the two-phase expanders significantly depends on the CO_2 stream. All three improvements concerning the refrigeration cycle side of the process are almost independent of the CO_2 stream composition: The effect of the aftercooler is almost equal for all considered CO_2 streams, both for the 2-stage and the 3-stage closed cycle process, and the benefit of using a phase separator differs for the 2-stage and the 3-stage process but is almost independent of the CO_2 stream. The smallest impact was found for the internal heat exchanger, which decreases the minimum specific energy demand of the 2-stage closed cycle process by 0.2 kWh/t CO_2 to 0.3 kWh/t CO_2 and by up to 0.1 kWh/t CO_2 for the 3-stage closed cycle process.

The efficiency of refrigeration cycles is usually quantified in terms of the COP, which relates the transferred heat (i.e. cooling power) to the required power. In contrast to a simple refrigeration cycle, the pressure of the feed CO_2 stream in the CO_2 closed cycle liquefaction process is reduced and heat is rejected at multiple temperature levels. Moreover, in the optimised closed cycle liquefaction processes, the enthalpy of the CO_2 feed stream is reduced by the CO_2 expanders, which in turn reduce the transferred heat in comparison to the closed cycle base processes. The COP of the closed cycle processes are therefore higher than the COP of conventional 2-stage or 3-stage ammonia refrigeration processes with a cooling temperature of -50 °C. Conventional 2-stage or 3-stage ammonia refrigeration processes are used for boil-off gas reliquefaction and consequently, lower COP have been calculated for boil-off gas reliquefaction (q. v. section 4.2). The COP of the 2-stage and the 3-stage closed cycle liquefaction processes are shown in Table 11. It can be seen that the highest COP were found for the optimised 3-stage closed cycle process while the lowest COP were found for the 2-stage closed cycle base process. However, in contrast to what the results in Figure 33 indicate, the COP is not generally highest for pure CO_2 and lowest for the Oxy96 CO_2 stream. This can be explained by the fact that both the electrical power input and the transferred heat depend on the composition of the

CO_2 stream. Thus, the COP calculated here does not only quantify the efficiency of the refrigeration cycles, but also takes the properties of the CO_2 stream into account. The COP in Table 11 therefore cannot be compared to the COP of conventional refrigeration cycles.

Table 11: Coefficients of performance (COP) of the 2-stage and the 3-stage closed cycle base and optimised processes for various CO_2 streams

Process design	Pure	Post	Oxy98	Oxy96	CL-MIN-CO2
2-stage base	2.56	2.56	2.57	2.58	2.55
2-stage optimised	3.89	3.87	3.49	3.33	3.89
3-stage base	2.95	2.94	2.96	2.97	2.96
3-stage optimised	4.60	4.57	4.17	4.01	4.14

In Figure 34, the individual components of the minimum specific energy demands of the optimised closed cycle processes are shown. The specific electrical energy demands of the refrigerant compressors are represented as a positive value and the electrical energy recovered by the expanders as negative ones. The summation of these values results in the respective minimum specific energy demands shown in Figure 33. For the 2-stage closed cycle process with pure CO_2, for example, the sum of the specific electrical energy demands of the refrigeration compressors is 12.3 kWh/t CO_2. From this value, 3.6 kWh/t CO_2 is deducted, which is the sum of the electrical energy recovered by the expanders, yielding the minimum specific energy demand of 8.7 kWh/t CO_2 shown in Figure 33.

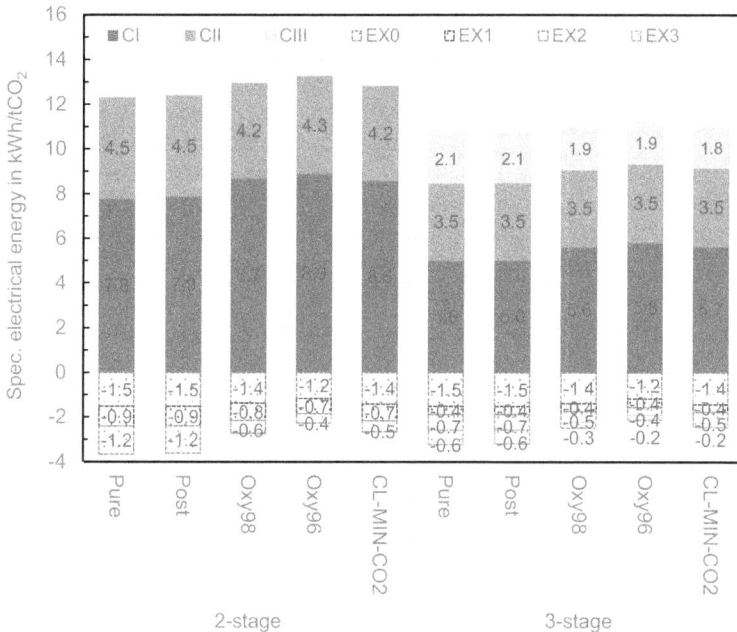

Figure 34: Specific electrical energy consumption (positive values) and recovery (negative values) in the optimised 2-stage and the optimised 3-stage closed cycle process. CI, CII, CIII are the refrigerant compressors, EX0 is the liquid CO_2 expander and EX1, EX2 and EX3 are the two-phase CO_2 expanders.

As mentioned above, the pressure ratio of the liquid expander depends on the CO_2 stream rather than on the number of stages. For this reason, the same energy reduction for a given CO_2 stream could be expected for the 2-stage and the 3-stage closed cycle process, but this is not the case. Moreover, according to the results shown in Figure 34, the electrical energy recovered is lower than the energy reduction obtained by installing the liquid expander (shown in Figure 33). These two observations suggest that installation of a liquid expander results not only in the recovery of electrical energy, but also reduces the energy demand of the refrigeration compressors. This can be attributed to the fact that the expander reduces the enthalpy of the CO_2 stream, whereas the pressure reduction via a valve does not change it significantly (ideally, an adiabatic isenthalpic expansion). In the case of the optimised 2-stage closed cycle process with pure CO_2, the temperature after expansion is lowered to 8.3 °C while it would be 9.9 °C without the liquid CO_2 expander. As a consequence, the cooling duty of the first CO_2/NH_3 heat exchanger (*HE1*) is 67 MJ/t CO_2 with a liquid CO_2 expander instead of 73 MJ/t CO_2 without one. This leads to a reduction in the minimum specific energy demand by 1.9 kWh/t CO_2 even though only 1.5 kWh/t CO_2 are recovered by the liquid CO_2 expander. The difference between the energy reduction for the optimised 2-stage and the optimised 3-stage closed cycle process can be explained by the higher efficiency of the 3-stage version: The energy demand reduction shown in Figure 33 is slightly lower for the optimised 3-stage closed cycle process as less electrical energy is required to meet the cooling demand.

While the pressure ratio of the liquid expander depends only to a small extent on the CO_2 stream, greater variations can be observed for the pressure ratios of the two-phase expanders. This applies especially to the last stage CO_2 expander, i.e. *EX2* for the optimised 2-stage closed cycle process and *EX3* for the optimised 3-stage version. In the last stage CO_2 expander, only a third of the energy is recovered in the case of the Oxy96 CO_2 stream with respect to pure CO_2. Taking into account the complexity of handling large variations in the pressure ratio, the technical feasibility and economic viability of installing a CO_2 expander for the last stage most likely depend on the variations in the impurity concentrations which are expected in the CO_2 stream. As for the liquid expander, the results for the two-phase expanders show that a considerable proportion of the energy reduction can be attributed to the reduction of the enthalpy during expansion. In the optimised 2-stage closed cycle process with pure CO_2, for example, the energy recovery by the two-phase expanders is 2.1 kWh/t CO_2 while the total reduction in the specific energy demand is 3.1 kWh/t CO_2.

It has not yet been discussed that the optimum values for the intermediate CO_2 pressures (p_1 and p_2) are different for the closed cycle base processes and the optimised versions. While the values given in Table 10 apply to the closed cycle base processes, the values for the optimised versions are generally higher. The higher intermediate CO_2 pressures of the optimised processes result in higher CO_2 temperatures after expansion and after condensation.

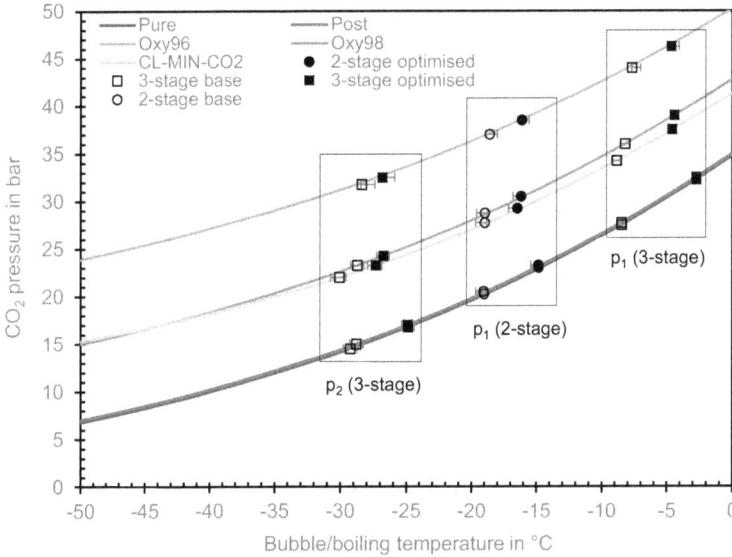

Figure 35: Optimum intermediate CO_2 pressures and the associated bubble/boiling temperatures for the optimised 2-stage and the optimised 3-stage closed cycle process. The coloured lines represent the bubble/boiling curves of the CO_2 streams.

The intermediate CO_2 pressures and bubble temperatures of the closed cycle base and optimised processes are shown in Figure 35. It can be seen that the differences between the base processes and the optimised versions in regard to the optimum intermediate CO_2 pressures and the resulting bubble temperatures decrease with increasing impurity concentration. The intermediate CO_2 pressures were varied by a step size of 0.25 bar to determine the optimum values for p_1 and p_2. The error bars in Figure 35 show the potential numerical imprecision caused by this step size. Depending on the CO_2 stream and CO_2 pressure, different values for the imprecision of the bubble temperature have been found. However, the imprecisions are rather small in absolute terms, so the differences concerning the optimum values for p_1 and p_2 cannot be attributed to numerical imprecisions.

Table 12: **Optimum intermediate pressures p_1 and p_2 of the optimised 2-stage and the optimised 3-stage closed cycle process and their impact on the minimum specific energy demand of the optimised process with respect to the optimum p_1 and p_2 of the respective base processes shown in Table 10**

CO_2 stream	Process design	p_1	p_2	p_3	Reduction of min. spec. energy demand in
Unit		bar	bar	bar	kWh/t CO_2
Pure	2-stage	23.00	6.77		0.08
	3-stage	32.25	16.75	6.77	0.12
Post	2-stage	23.25	7.03		0.11
	3-stage	32.50	17.00	7.03	0.11
Oxy98	2-stage	30.50	15.03		0.05
	3-stage	39.00	24.25	15.03	0.06
Oxy96	2-stage	38.50	23.71		0.04
	3-stage	46.25	32.50	23.71	0.04
CL-MIN-CO2	2-stage	29.25	15.40		0.04
	3-stage	37.50	23.25	15.40	0.07

The results in Figure 35 raise the question of whether the differences between the optimum p_1 and p_2 of the closed cycle base processes and the optimised versions have a significant influence on the minimum specific energy demand. In Table 12, the optimum intermediate CO_2 pressures of the optimised 2-stage and the optimised 3-stage closed cycle process are shown. The values in the last column show to what extent the minimum specific energy demand is influenced by the change in the optimum intermediate pressures. To obtain the values in the last column, the minimum specific energy demand of the optimised closed cycle process was calculated using the optimum p_1 and p_2 of the *base* process shown in Table 10. From these values, the minimum specific energy demand of the optimised closed cycle process using the optimum p_1 and p_2 of the *optimised* process (i. e. the values shown in Figure 35) is deducted.

The results in Table 12 show that the reduction in the minimum specific energy demand caused by the adaptation of the intermediate CO_2 pressures is comparably small. For example, the minimum specific energy demand of the optimised 2-stage closed cycle process can be reduced by 0.08 kWh/t CO_2 when an intermediate CO_2 pressure of 23 bar (optimum value of the optimised process) instead of 20.25 bar (optimum value of the base process) is selected. This small difference in the energy demand shows that the intermediate CO_2 pressures do not necessarily need to be adapted for the optimised closed cycle processes. This represents a significant advantage for practical implementation since it allows the retrofit of individual components of the optimised process (e.g. CO_2 expanders or ammonia aftercooler) with only a slight or no change in the intermediate CO_2 pressures.

4.1.4 CO_2 Feed Stream and Ambient Conditions

In all previous results on liquefaction, a CO_2 feed pressure of 100 bar and a CO_2 feed temperature of 15 °C as well as a seawater temperature of 15 °C have been assumed. Since in reality, these parameters depend on the CO_2 pipeline and ambient conditions, the impact on the minimum specific energy demand has been studied.

In Figure 36, the minimum specific energy demands for the liquefaction of pure CO_2 are shown in dependency of the CO_2 feed pressure for the 2-stage and the 3-stage closed cycle base process as well as for the optimised versions. The results in Figure 36 show a similar dependency from the CO_2 feed pressure for the two base processes as well as for the two optimised processes. For the two base processes, the dependency of the minimum specific energy demand from the CO_2 feed pressure is rather low. This can be explained by the fact that the CO_2 stream pressure is reduced by a valve, so energy is not recovered. The slightly lower specific energy demand for high feed pressures can be attributed to the Joule-Thompson effect, which leads to lower temperatures after expansion for higher CO_2 feed pressures.

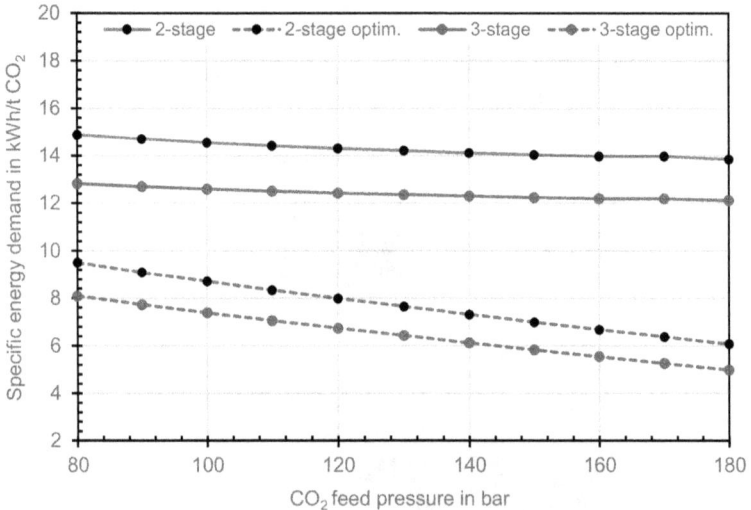

Figure 36: **Minimum specific energy demand for the liquefaction of pure CO_2 in dependency of the CO_2 feed pressure**

The reduction of the energy demand is much higher for the optimised processes, as more energy is recovered by the CO_2 expanders for higher CO_2 feed pressures. Therefore, the minimum specific energy demand of the optimised 3-stage closed cycle process is approximately 0.7 kWh/t CO_2 (9 %) lower for a CO_2 feed pressure of 120 bar than in the reference case of 100 bar. On the other hand, the higher CO_2 feed pressure leads to a higher energy demand from the CO_2 compressor at the CCS plant. Therefore, increasing the CO_2 pipeline pressure is not a sensible measure per se.

In Figure 37, the minimum specific energy demands for the liquefaction of pure CO_2 are shown in dependency of the CO_2 feed temperature for the 2-stage and the 3-stage closed cycle base process as well as the optimised versions. An increase of the minimum specific energy demand for an increasing CO_2 feed temperature can be observed for all considered processes. This can be explained by the fact that the enthalpy of the CO_2 stream, and hence, the cooling duty, increases with an increasing CO_2 feed temperature. The highest influence of the CO_2 feed temperature was found for the 2-stage closed cycle base process, while the lowest was found for the optimised 3-stage closed cycle process.

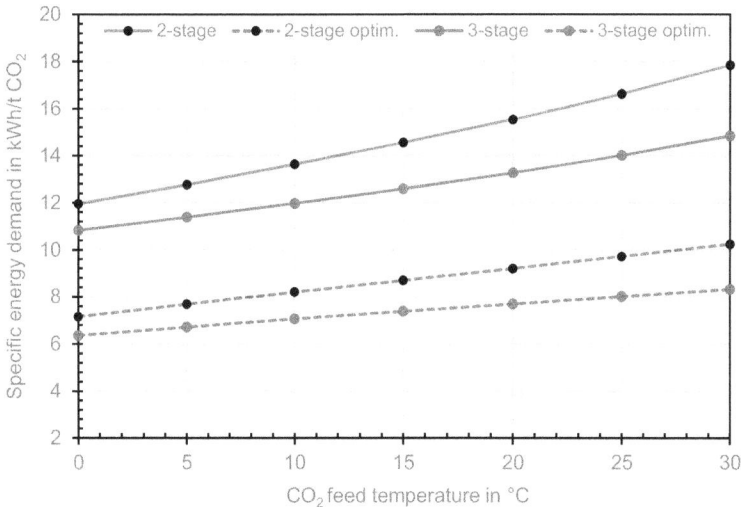

Figure 37:   Minimum specific energy demand for the liquefaction of pure CO_2 in dependency of the CO_2 feed temperature

The differences in the influence of the CO_2 feed temperature can be attributed to the different efficiencies of these processes: As the 2-stage closed cycle base process has the lowest efficiency and the optimised 3-stage closed cycle process has the highest among all processes considered, an increase in the cooling duty has the highest impact on the minimum specific energy demand of the 2-stage closed cycle base process and the lowest impact on the optimised 3-stage closed cycle process. Thus, the doubling of the CO_2 feed temperature from 15 °C to 30 °C leads to an increase in the minimum specific energy demand by approximately 23 % for the 2-stage closed cycle base process yet only 13 % for the optimised 3-stage closed cycle process.

In Figure 38, the minimum specific energy demands for the liquefaction of pure CO_2 are shown in dependency of the seawater temperature for the 2-stage and the 3-stage closed cycle base process as well as for the optimised versions. For all considered processes, an increase in the minimum specific energy demand can be seen with increased seawater (i. e. cooling water) temperature.

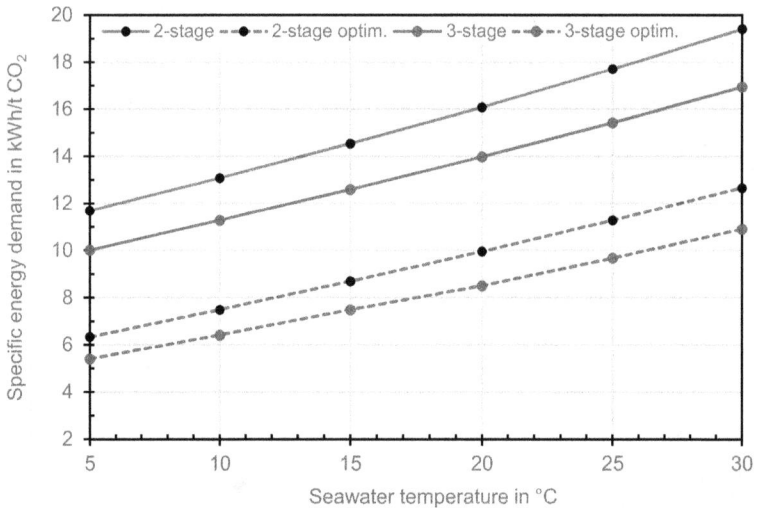

Figure 38:   Minimum specific energy demand for the liquefaction of pure CO_2 in dependency of the seawater (i.e. cooling water) temperature

In comparison to Figure 37, the curves in Figure 38 do not diverge as much for higher seawater temperatures. This means that the impact of the liquefaction process on the minimum specific energy demand is smaller for the seawater temperature than for the CO_2 feed temperature. In contrast to Figure 37, the gradients of the optimised processes are even slightly higher than the gradients of the base processes. The impact of the seawater temperature is generally higher than the impact of the CO_2 feed temperatures: By doubling the seawater temperature from 15 °C to 30 °C, the minimum specific energy demand is increased by approximately 34 % for the 2-stage closed cycle base process while it is increased by approximately 46 % for the optimised 3-stage closed cycle process.

4.1.5 Liquefaction Temperature

So far, a transport temperature and consequently, a liquefaction temperature of -50 °C has been assumed for the reasons stated in section 3.1. A higher liquefaction temperature would decrease the energy demand for liquefaction at the expense of a higher transport pressure. In Figure 39, the minimum specific energy demands for the liquefaction of pure CO_2 are shown in dependency of the liquefaction temperature for the 2-stage and the 3-stage closed cycle base process as well as the optimised versions.

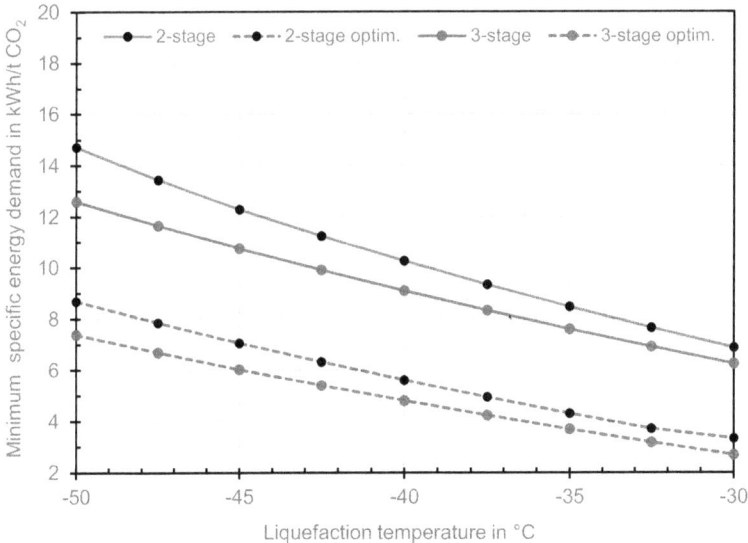

Figure 39: Minimum specific energy demand for the liquefaction of pure CO_2 in dependency of the liquefaction temperature

85

For an increasing liquefaction temperature, a significant decrease in the minimum specific energy demand can be seen for all considered processes. The strongest decrease can be observed for the optimised processes: For example, the minimum specific energy demand of the 2-stage closed cycle base process is approximately 30 % lower when the liquefaction temperature is increased from -50 °C to -40 °C while it is 35 % lower for the optimised 2-stage closed cycle process. Similarly, the decrease is 28 % for the 3-stage closed cycle base process while it is 35 % for the optimised 3-stage closed cycle process. For a liquefaction temperature of -30 °C, reductions in the minimum specific energy demand between 50 % (3-stage closed cycle base process) and 64 % (optimised 3-stage closed cycle process) have been determined. The impact of an additional liquefaction temperature increase decreases with increased liquefaction temperatures: While an increase from -50 C to -45 °C leads to a reduction by approximately 15 % to 18 %, the reduction is only between 11 % and 14 % for an increase from -35 °C to -30 °C. On the other hand, a decrease in the liquefaction temperature leads to an increase in the liquefaction pressure (q. v. section 3.1), and consequently, to an increase in the design pressure for all components of the ship transport chain, most importantly those of the onshore intermediate storage and ship tanks. For this reason, most authors propose a liquefaction temperature of-50 °C for large-scale CO_2 transport (q. v. section 3.1), at the expense of a higher energy demand.

4.2 Boil-Off Gas Handling

Three strategies are possible for boil-off gas: It can be reliquefied using a refrigeration process, it can be retained inside the tank, accepting the pressure increase ("no venting"), or it can be released into the atmosphere.

The main difference between boil-off gas reliquefaction and the liquefaction of a pipeline CO_2 stream is that heat is rejected at transport temperature and transport pressure only (e.g. -50 °C and 6.75 bar) in boil-off gas reliquefaction. The refrigeration cycle is therefore independent of the CO_2 stream properties. A sensitivity analysis has been carried out to determine the optimum operating parameters for the 2-stage and the 3-stage closed cycle reliquefaction processes. The results show that the optimum lower temperature cycle compressor ratio is 4.8 regardless of whether a 2-stage or 3-stage closed cycle process is used. This is a significant difference with the pipeline CO_2 liquefaction process, where the pressure ratios of the 3-stage closed cycle process are generally lower. Consequently, the difference between the efficiency of the 2-stage closed cycle process and the efficiency of the 3-stage closed cycle process is lower in the case of boil-off gas reliquefaction. The optimum pressure ratios for the refrigeration cycle compressors are given in Table 13. The maximum COP was found to be 2.01 for the 2-stage process and 2.16 for the 3-stage process. With these values, the specific electrical energy demand necessary for boil-off gas reliquefaction was calculated for each CO_2 stream.

Table 13: Optimum pressure ratios of the refrigeration cycle compressors for the boil-off gas reliquefaction processes

	Π_{CI}	Π_{CII}	Π_{CIII}
2-stage	5.3	4.8	
3-stage	2.0	2.6	4.8

Figure 40 shows the specific electrical energy demand for the reliquefaction of boil-off gas from various CO_2 streams as well as the enthalpy of evaporation of the boil-off gas. Compared to the liquefaction of a pipeline CO_2 stream, the difference between the individual streams is much higher for boil-off gas reliquefaction. For example, the specific energy demand for the reliquefaction of boil-off gas from pure CO_2 and the CL-MIN-CO2 CO_2 stream differs by a factor of approximately 13. The significant difference between these CO_2 streams can be explained by the different enthalpies of evaporation: With a higher fraction of volatile components such as oxygen, argon or hydrogen, the enthalpy of evaporation rises. Thus, more heat needs to be transferred to condense the boil-off gas stream, leading to a higher specific electrical energy demand. The higher enthalpy of evaporation for the CL-MIN-CO2 CO_2 stream can mainly be explained by the higher fraction of hydrogen, which has the most significant influence on the fluid properties (and thus, on the enthalpy of evaporation) among all considered CO_2 impurities. The boil-off gas composition for various CO_2 streams is shown in Figure 41. It can be seen that the fraction of CO_2 in the CO_2 stream is approximately 50 % for the Oxy98 and the CL-MIN-CO2 CO_2 streams.

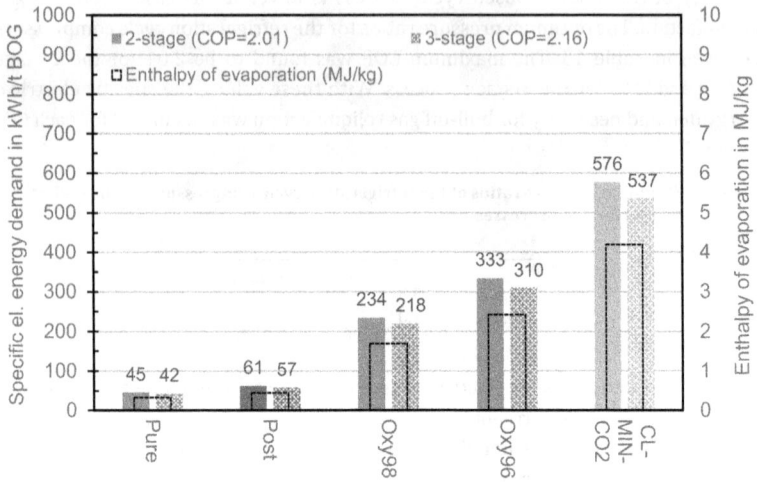

Figure 40: Specific electrical energy demand for boil-off gas reliquefaction from various CO₂ streams and the respective enthalpy of evaporation of the boil-off gas

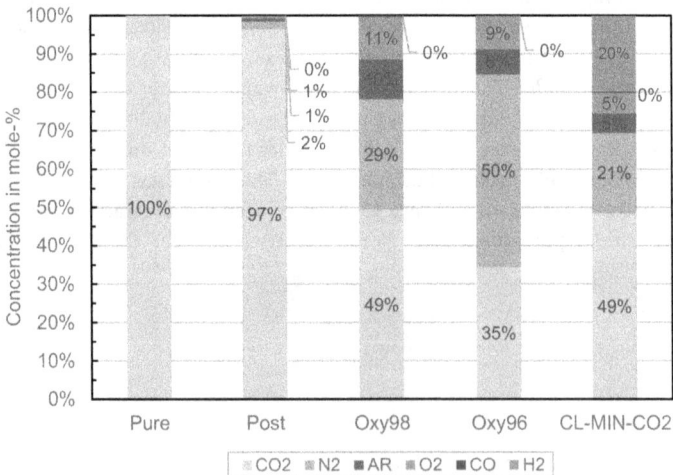

Figure 41: Boil-off gas composition for various CO₂ streams

The results in Figure 40 and Figure 41 suggest that boil-off gas reliquefaction becomes increasingly inefficient with increased impurity concentrations: Firstly, the energy demand increases (Figure 40). Secondly, the concentration of CO_2 decreases with an increased impurity concentration and is generally below 50 % (with the exception of the Post boil-off gas stream). Therefore, the question arises as to whether the boil-off gas could be retained in the CO_2 tank during transport as an alternative, accepting the increase in pressure ("no venting"), or if the boil-off gas could be released into the atmosphere. These two alternative options are discussed in section 4.5 on the basis of three different example scenarios.

4.3 Injection Process

The injection process consists of the pressure increase from tank pressure to wellhead pressure and the subsequent heating of the CO_2 stream from -50 °C to 5°C. Figure 42 shows the specific electrical energy demand of the injection pump and the specific thermal energy demand required for CO_2 heating as a function of the wellhead pressure for various CO_2 streams. The electrical energy demand increases linearly with increasing wellhead pressure as both are proportional to the pressure ratio. The pressure ratio is linearly dependent on the wellhead pressure as the input pressure (i.e. the tank pressure) is determined by the composition of the CO_2 stream which is the same for all wellhead pressures of every CO_2 stream quality. With an increased impurity concentration, the electrical energy demand decreases due to the increased tank pressure. At the same time, the specific thermal energy demand for CO_2 heating decreases for higher wellhead pressures and increases for an increased impurity concentration. This can be explained by the fact that the output temperature of the injection pump increases with an increasing pressure ratio.

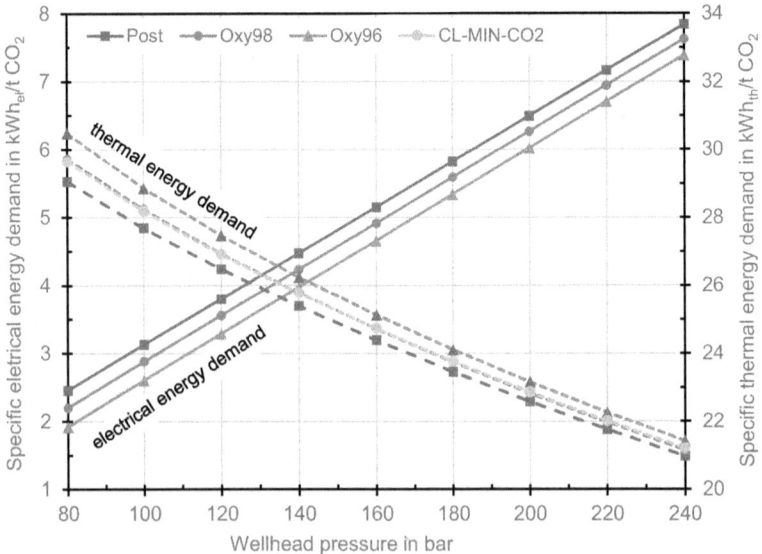

Figure 42: Specific electrical and thermal energy demand for CO_2 injection for various CO_2 stream compositions. The results for pure CO_2 are not shown as they overlap with the results for the Post CO_2 stream.

Figure 43 illustrates in a Q, T diagram how the thermal energy demand for pure CO_2 is provided by seawater heat, engine waste heat and additional heat in dependency of the wellhead pressure. A seawater temperature of 4 °C is assumed, as it leads to a maximum potential additional heat demand. The different temperatures in the CO_2 at 0 MJ/t CO_2 can be explained by the different outlet temperatures of the injection pump due to differences in the wellhead pressure. With the assumed upper minimum internal temperature approach of 5 K, the CO_2 can be heated up to -1 °C by seawater. The engine waste heat streams (lower and upper temperature cooling cycle and the exhaust gas stream) are used to further increase the temperature. Only for a wellhead pressure of 240 bar are seawater heat and engine waste heat sufficient to reach the desired temperature of 5 °C. For lower wellhead pressures, additional heat must be provided.

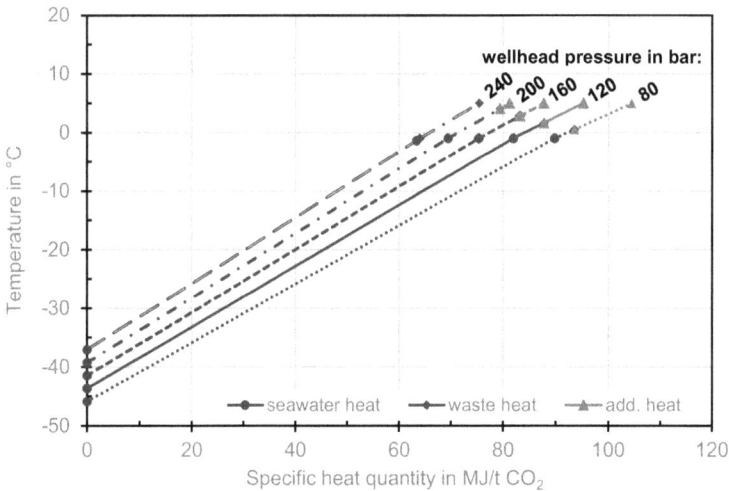

Figure 43: Q, T diagram for the heating of pure CO_2 for various wellhead pressures. A seawater temperature of 4 °C is assumed

Figure 44 shows under which conditions additional heat must be provided to meet the thermal energy demand according to Figure 42. The results for pure CO_2 can be found in another publication [95]. While the total thermal energy demand depends only on the wellhead pressure and the CO_2 stream composition, the additional heat demand also depends on the seawater temperature. It can be seen that the additional heat demand is generally higher for the CO_2 streams with impurities than for a pure CO_2 stream. This can be explained by the lower outlet temperature of the injection pump caused by the lower compression ratio.

Figure 44 shows that additional heat is not required for a seawater temperature of 10 °C or higher. Even at 8 °C, no additional heat is required when the wellhead pressure is at least 100 bar. Conversely, additional heat must be provided at a seawater temperature of 4 °C for all wellhead pressures between 80 bar and 220 bar, except for the Oxy96 CO_2 stream where additional heat must also be provided for a wellhead pressure of 240 bar. In most cases, the additional heat demand is rather low in comparison to the total thermal energy demand. For example, the additional energy demand at a wellhead pressure of 120 bar and a seawater temperature of 6 °C is 3.3 % of the total thermal energy demand for the Post CO_2 stream and 3.8 % for the Oxy98 CO_2 stream.

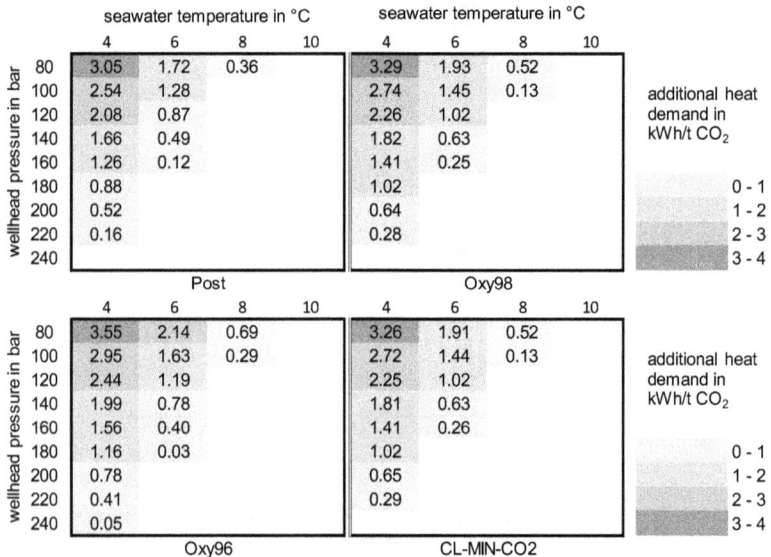

Post — seawater temperature in °C

wellhead pressure in bar	4	6	8	10
80	3.05	1.72	0.36	
100	2.54	1.28		
120	2.08	0.87		
140	1.66	0.49		
160	1.26	0.12		
180	0.88			
200	0.52			
220	0.16			
240				

Oxy98 — seawater temperature in °C

wellhead pressure in bar	4	6	8	10
80	3.29	1.93	0.52	
100	2.74	1.45	0.13	
120	2.26	1.02		
140	1.82	0.63		
160	1.41	0.25		
180	1.02			
200	0.64			
220	0.28			
240				

additional heat demand in kWh/t CO_2

- 0 - 1
- 1 - 2
- 2 - 3
- 3 - 4

Oxy96

wellhead pressure in bar	4	6	8	10
80	3.55	2.14	0.69	
100	2.95	1.63	0.29	
120	2.44	1.19		
140	1.99	0.78		
160	1.56	0.40		
180	1.16	0.03		
200	0.78			
220	0.41			
240	0.05			

CL-MIN-CO2

wellhead pressure in bar	4	6	8	10
80	3.26	1.91	0.52	
100	2.72	1.44	0.13	
120	2.25	1.02		
140	1.81	0.63		
160	1.41	0.26		
180	1.02			
200	0.65			
220	0.29			
240				

additional heat demand in kWh/t CO_2

- 0 - 1
- 1 - 2
- 2 - 3
- 3 - 4

Figure 44: Additional heat demand for injection as a function of the wellhead pressure and the seawater temperature

4.4 Transport Chain Model

4.4.1 Baseline Scenarios

Three baseline scenarios were defined to analyse the impact of different CO_2 quantities and impurity concentrations on the individual components of the transport chain. A distance of 100 km and a velocity of 15 knots are assumed in accordance with the assumptions in the CLUSTER project. In addition to these baseline scenarios, the impact of the distance is studied in section 4.4.3 by the example of a 1000 km scenario, which roughly correlates with the distance from the German North Sea coast to the Norwegian Sleipner CO_2 storage location (Utsira Sand saline aquifer [96]). For all baseline scenarios, a loading time of 0.5 h is assumed and the loading mass flow rate is selected accordingly. The injection mass flow rate is assumed to be 20 % higher than the respective source mass flow rate. Additional handling times for manoeuvring are assumed for loading and unloading (or injection), respectively. Two ships are used in all three baseline scenarios.

In the first scenario, a pure CO_2 stream with a constant mass flow rate and a total quantity of 1 Mt/a are assumed. The 2 Mt/a - Oxy96 and the 20 Mt/a - Cluster scenarios are based on the feed-in characteristics of typical power and industrial plants. Contrary to the first scenario, variable CO_2 mass flow rates are assumed. The feed-in characteristics have been defined within the scope of the joint research project CLUSTER and are described in detail in another Ph.D. thesis [70]. The second scenario is based on the "SK-Oxy" power plant defined in the CLUSTER project's baseline scenario, which is a hard-coal fired power plant with Oxyfuel combustion capture technology. Its net power output is 446.7 MW and the net efficiency is 36.6 %. A gas processing unit (GPU) is used to attain a CO_2 purity of 96 vol.-% (Oxy96 CO_2 stream). With a full load mass flow rate of 390 t/h and typical medium load characteristics, the total quantity of CO_2 per year fed into the pipeline is approximately 2.3 Mt/a, including impurities. The CO_2 production characteristic of the power plant is shown in the appendix (section A.2, Figure A 2). The third scenario is based on a cluster of typical CO_2 emitters that consists of power plants, cement and steel plants as well as refineries. The three major CO_2 capture technologies, Post-Combustion, Pre-Combustion and Oxyfuel, are used. With a full load mass flow rate of 3236 t/h, the total annual quantity of CO_2 fed into the pipeline is approximately 19.8 Mt/a, which includes impurities. The CO_2 production characteristic of the cluster is shown in the appendix (section A.2, Figure A 3). The composition of the CO_2 stream depends on the mass flow rates and capture technologies of the individual plants that contribute to the total pipeline CO_2 stream. The CO_2 composition CLUSTER-MIN-CO2 is assumed to ensure conservative estimates for the onshore intermediate storage and ship capacity as well as for the energy demand.

The main properties of the three baseline scenarios are shown in Table 14. The design mass flow rate is defined as the maximum mass flow rate that occurs in the pipeline to ensure that the entire CO_2 stream can be transported at any given time. The transport chain capacity per year is equal to the design mass flow rate, merely given in the unit Mt/a. In all three scenarios, the transport chain capacity per year is greater than the transported quantity per year (which is either 1 Mt/a, 2.3 Mt/a or 19.8 Mt/a). While only slightly greater for the 1 Mt/a – Pure CO_2 scenario, it is considerably higher for the other two scenarios. This difference follows from the fact that the transport chain is designed for the full load mass flow rate, whereas the transported quantity per year is determined by the actual feed-in of the power plant or the CO_2 emission cluster. The total number of roundtrips (cumulated for both ships) depends on the roundtrip time, the number of ships and the ratio of the transported quantity to the transport chain capacity. The total number of roundtrips is the highest for the 2 Mt/a - Oxy96 scenario primarily because it has the lowest roundtrip time. While the roundtrip time is the same for the 1 Mt/a – Pure CO_2 scenario and the 20 Mt/a - Cluster scenario, the ratios of the transported quantity to the transport chain capacity are not equal. The full transport chain capacity is utilised in the 1 Mt/a – Pure CO_2 scenario (i.e. the ratio is 1) and thus, a higher number of roundtrips is necessary than in the 20 Mt/a - Cluster scenario.

Table 14: Main properties of the three baseline scenarios

Category	Parameter	Unit	1 Mt/a - Pure CO₂	2 Mt/a - Oxy96	20 Mt/a - Cluster
General	Design mass flow rate	t/h	114	390	3236
	Transport chain capacity per year	Mt/a	1.0	3.42	28.35
	Transported quantity per year	Mt/a	1.0	2.3	19.8
	Roundtrip time	h	18.3	11.2	18.3
	Total no. of roundtrips per year	-	955.2	1064.6	666.6
	Liquid CO₂ density	t/m³	1.154	1.143	1.147
	Vaporous CO₂ density	t/m³	0.018	0.034	0.028
Inter-mediate	Gross onshore intermediate storage capacity	t	1005	2052	28774
Loading	Loading mass flow rate per tank	t/h	2094	1093	2968
	Loading time	h	0.5	0.5	0.5
	Handling time	h	1	1	1
Ship	Gross tank capacity	t	1063	563	1522
	Number of tanks	-	1	4	20
	Gross ship capacity	t	1063	2253	30434
	Number of Ships	-	2	2	2
	Net total capacity of ships	t	2094	4372	59358
Transport	Distance	km	100	100	100
	Speed	kn	15	15	15
	Transport time	h	3.6	3.6	3.6
Unloading	Unloading mass flow rate per tank	t/h	137	1093	194
	Unloading time	h	7.6	0.5	7.6
	Handling time	h	0.5	0.5	0.5
	Gross offshore intermediate storage capacity	t	0	2011	0
Injection	Injection mass flow rate per well	t/h	137	234	431
	Number of wells	-	1	2	9
	Injection time	h	7.6	4.5	7.6

The gross ship capacity is determined by the gross tank capacity and the number of tanks. For the gross tank capacity, the limitations described in section 2.3 are considered: A design pressure of 9 bar is assumed for the pure CO₂ stream (vapour pressure of approximately 7 bar plus a maximum $(p_{gd})_{max}$ of 2 bar), a design pressure of 26 bar for the Oxy96 CO₂ stream and a design pressure of 17 bar for the CL-MIN-CO2 CO₂ stream. Thus, the maximum volume for a tank with a length of 50 m is approximately 4800 m³ in the case of pure CO₂ (which corresponds with a gross capacity of 5541 t), 600 m³ (686 t) for the Oxy96 CO₂ stream and 1350 m³ (1549 t) for the CL-MIN-CO2 CO_2 stream when assuming an allowable membrane pressure of 200 N/mm^2 (q.v. Figure 10). Consequently, only one tank is necessary for the 1 Mt/ - Pure CO_2 scenario since the gross tank capacity is 1063 t. For the 2 Mt/a - Oxy96 scenario, four tanks with a gross capacity of 563 t are selected. A two-tank configuration would lead to a tank capacity above the limit of 686 t and was therefore not chosen, although it might be selected in practice since the calculations in Chapter 2 represent only an estimation. A three tank configuration was not selected as an even number of tanks provides a better use of the available space on the ship. For the 20 Mt/a - Cluster scenario, a gross tank capacity of 1522 t is assumed, so that 20 tanks are necessary to reach the calculated gross ship capacity of 30434 t. The number of tanks could be reduced from 20 to 10 by increasing the tank length from 50 m to 100 m. In this case, the loading mass flow rate per tank must be doubled to keep the overall transport chain capacity constant. Therefore, the tank capacity, number of tanks and mass flow rate per tank must be considered in conjunction. This means that the values given in Table 14 represent just one of the possible solutions for each scenario.

Comparing the dimensions of the transport chain components in Table 14 (e.g. the onshore intermediate storage and ship capacities), it can be seen that the values for the 20 Mt/a - Cluster scenario are greater by a factor of approximately 28 with respect to the 1 Mt/a - Pure CO_2 scenario, which corresponds to the difference in the transport chain capacities of these scenarios. The net total capacity of ships differs by a factor of 28.35 (which corresponds to the ratio of the transport chain capacities), and the gross ship and gross onshore intermediate storage capacities differ by factor of 28.6 due to the different vaporous and liquid densities. Thus, the 20 Mt/a - Cluster scenario merely represents a scaled-up version of the 1 Mt/a - Pure CO_2 scenario. Although both scenarios differ not only in the design capacity, but also in the CO_2 stream composition and CO_2 production characteristics (constant mass flow rate or varying mass flow rate), only the transported quantity is relevant for the transport chain schedule and the net capacity of its components. However, the different CO_2 stream compositions and CO_2 production characteristics have an impact on the energy demand of the overall transport chain, which is discussed in section 4.5.

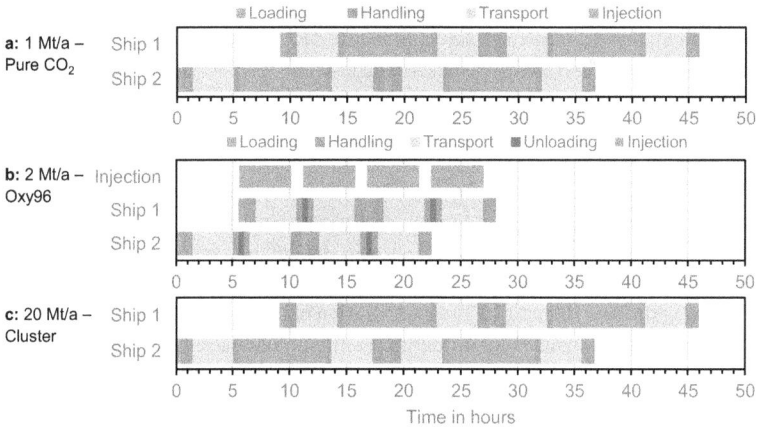

Figure 45: Schedule of the three baseline scenarios presented in Table 14. Two roundtrips are shown for each ship.

A graphical representation of the transport chain schedules is shown in Figure 45. For reasons of better graphical readability, only two roundtrip cycles are displayed. It can be seen that the 1 Mt/a – Pure CO_2 scenario and the 20 Mt/a - Cluster scenario have the same schedule, since the roundtrip time and all of its components are equal. The impact of uncoupling the unloading and injection procedure is illustrated in the 2 Mt/a - Oxy96 scenario. By using offshore intermediate storage, the unloading time can be reduced independently of the injection mass flow rate. In the 2 Mt/a - Oxy96 scenario, the unloading mass flow rate is selected so that the unloading time is equal to the loading time. An effect of this is that the roundtrip time is shorter than in the other two scenarios, which leads to a higher transport chain capacity per year for a given ship capacity. Conversely, for a given transport chain capacity per year, the ship capacity is lower. This can be observed when comparing the 1 Mt/a - Pure CO_2 and the 2 Mt/a - Oxy96 scenarios: While the gross ship capacity of 2253 t in the 2 Mt/a - Oxy96 scenario is only approximately twice the capacity of the 1 Mt/a - Pure CO_2 scenario, the transport chain capacity per year is increased by a factor of 3.4. On the other hand, gross offshore intermediate storage with a capacity of 2011 t is necessary, which corresponds to a barge with almost the same capacity as the transport ships. The ultimate decision between a larger ship capacity and using offshore intermediate storage must be based on an economic assessment.

97

The results of a sensitivity analysis for the 1 Mt/a - Pure CO_2 scenario and the 20 Mt/a - Cluster scenario are shown in Figure 46. The parameters are varied with a factor between 1 and 5 with respect to the baseline scenario and the impact on the transport chain capacity per year is studied. It has to be noted that varying a single parameter generally means that not all conditions stated in section 3.5 can be fulfilled. For the results shown in Figure 46, this is neglected as the impact of a single parameter shall be illustrated. The results show that the number of ships directly corresponds with the transport chain capacity per year. This can be explained by the fact that number of ships is a linear factor in the defining equation eq. (3.19). The other parameters shown in Figure 46 merely contribute to it indirectly by influencing the roundtrip time. The contribution of the individual component to the roundtrip time determines its influence: As the unloading time and the transport time represent the largest fractions of the roundtrip time in the 1 Mt/a – Pure CO_2 scenario and the 20 Mt/a - Cluster scenario, the unloading mass flow rate, distance and velocity influence the transport chain capacity per year significantly while the loading mass flow rate has the lowest influence. The ship capacity has an impact on both the loading time and the unloading time. In particular its impact on the unloading time explains the significant influence on the transport chain capacity per year. As in reality, the individual parameters cannot be changed independently, the potential increase in the transport chain capacity is generally lower than in Figure 46. Thus, a more comprehensive sensitivity analysis is carried out in the next two sections.

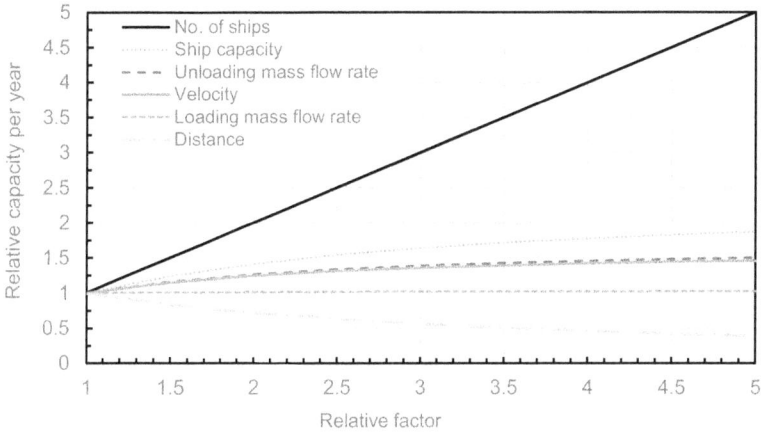

Figure 46: Sensitivity analysis for the 1 Mt/a - Pure CO_2 scenario and the 20 Mt/a - Cluster scenario. The results represent relative values with respect to the respective baseline scenarios shown in Table 14.

4.4.2 Number of Ships and Ship Capacity

In Figure 47, the impact of the number of ships on the transport chain schedule is shown for the 20 Mt/a - Cluster scenario. The remaining parameters are adjusted so that the conditions stated in section 3.5 are fulfilled. As shown in section 4.4.1, the same schedule is also valid for the 1 Mt/a - Pure CO_2 scenario. Moreover, the general considerations also apply to the 2 Mt/a - Oxy96 scenario. The 20 Mt/a - Cluster scenario was selected since the onshore intermediate storage and ship capacities would be very low in the other two scenarios for the assumed transport distance of 100 km.

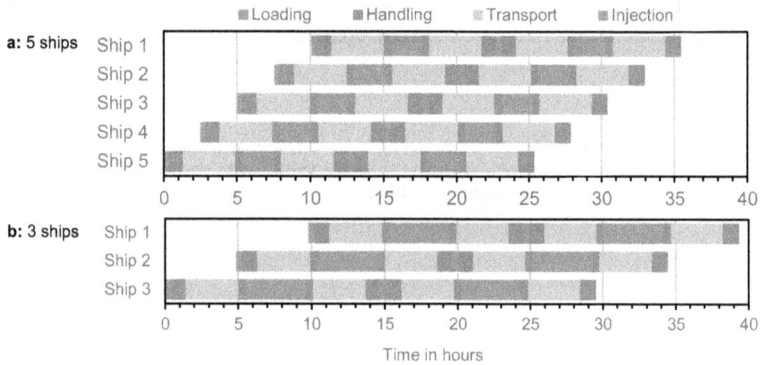

Figure 47:   Transport chain schedule for the 20 Mt/a - Cluster scenario when using 5 or 3 ships

Table 15:   Results for the 20 Mt/a - Cluster scenario when using 5, 3 or 2 ships

Parameter	Unit	5 ships	3 ships	2 ships
Roundtrip time	h	12.7	14.7	18.3
Total number of roundtrips per year	-	2415.9	1244.3	666.6
Gross onshore intermediate storage capacity	t	7253	14823	28774
Loading time	h	0.3	0.4	0.5
Gross tank capacity	t	1050	1359	1522
Number of tanks	-	8	12	20
Gross ship capacity	t	8398	16305	30434
Net total capacity of ships	t	40946	47701	59358
Unloading and Injection time	h	2.1	4.1	7.6

As can be seen from Figure 47, the loading and injection times per roundtrip decrease with an increased number of ships, thus the roundtrip time also decreases. The individual values are shown in Table 15: Since the same loading mass flow rate per tank (2968 t/h) and the same injection mass flow rate (431 t/h per well) as in the base scenario (q. v. Table 14) are used, the loading time and the injection time (per roundtrip) are reduced by using three or five instead of two ships. On the other hand, the number of roundtrips, both per ship and in total, increases with an increased number of ships. Since the reduction of the loading time is small in comparison to the increase of the number of roundtrips, the cumulated loading time, both per ship and in total, increases. The cumulated injection time, in contrast, is the same for all scenarios shown in Table 15 since the injection mass flow rates are equal and consequently, the increase of the number of roundtrips is compensated by the decrease of the injection time (per roundtrip).

The increase in the total number of roundtrips per year directly corresponds with the decrease in the ship capacity and, indirectly, with the decrease in onshore intermediate storage capacity: While 1.9 times more roundtrips are required for the 3 ships scenario with respect to the 2 ships scenario, the gross ship capacity is decreased by 46 % in turn. As a result, the cumulated, net total capacity of the ships is decreased by 20 %. The gross onshore intermediate storage capacity is reduced by 48 %. The decrease in the onshore intermediate storage capacity is generally higher than the decrease of the ship capacity since the time period between one ship leaving and the arrival of the next one, which is relevant for onshore intermediate storage capacity (q. v. eq. (3.11)), decreases at a slightly higher rate than the roundtrip time, which is relevant for the gross ship capacity. Similar values are found for the comparison of the 5 ships scenario to the 3 ships scenario: The gross ship capacity is reduced by 48 % and the gross intermediate storage capacity is reduced by 51 %.

4.4.3 Impact of Transport Distance

While a transport distance of 100 km has been assumed for the baseline scenarios, potential offshore CO_2 storage locations such as the Sleipner field are typically located further away from the coast. The impact of distance is studied using the example of the 20 Mt/a - Cluster scenario. Since the potential benefit of using an offshore intermediate storage is also evaluated, the results can be transferred to both the 1 Mt/a - Pure CO_2 scenario and the 2 Mt/a - Oxy96 scenario.

101

Figure 48 shows the schedule for a transport distance of 1000 km when using either 2 ships, 3 ships or 3 ships with offshore intermediate storage. It can be seen that the transport time – and therefore the roundtrip time – is significantly higher for a transport distance of 1000 km than for a transport distance of 100 km. As a consequence, the total number of trips per year is lower and the ship and intermediate storage capacities must be increased.

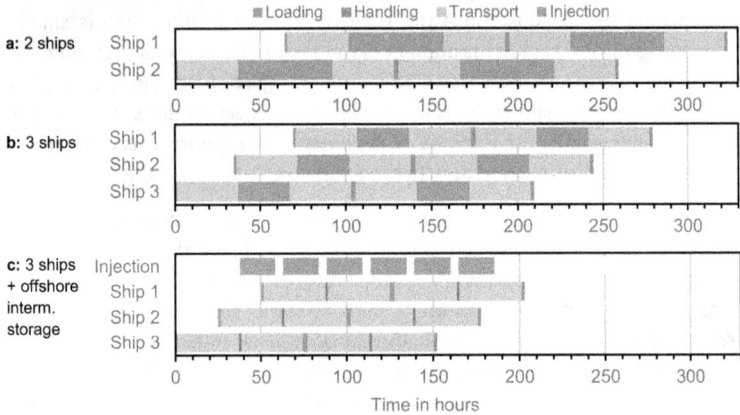

Figure 48: Transport chain schedule for a transport distance of 1000 km when using 2, 3 or 3 ships with offshore intermediate storage

Table 16: Results for the 20 Mt/a - Cluster scenario and a transport distance of 1000 km when using 2, 3, 5 or 3 and 5 ships with an offshore intermediate storage (o.i.st.)

Parameter	Unit	5 ships + o.i.st.	3 ships + o.i.st.	5 ships	3 ships	2 ships
Roundtrip time	h	76.0	76.0	90.6	104.5	129.4
Total number of roundtrips per year	-	402.3	241.4	337.5	175.5	94.5
Gross onshore intermediate storage capacity	t	48809	82411	58491	113974	213071
Loading time	h	0.5	0.5	0.5	0.5	0.5
Gross tank capacity	t	1483	1501	1503	1521	1534
Number of tanks	-	34	56	40	76	140
Gross ship capacity	t	50419	84039	60120	115626	214732
Net total capacity of ships	t	245839	245863	293143	338274	418810
Unloading time	h	0.5	0.5	15.1	29.0	53.9
Gross offshore intermediate storage capacity	t	48494	82091	0	0	0
Injection time	h	12.7	21.1	15.1	29.0	53.9

The results for the 20 Mt/a - Cluster scenario for a transport distance of 1000 km are shown in Table 16: Similar to a transport distance of 100 km, the gross ship capacity can be reduced by 46 % and the gross onshore intermediate storage capacity by 47 % when using 3 instead of 2 ships. Likewise, when using 5 instead of 3 ships, the gross ship capacity is reduced by 46 % and the gross onshore intermediate storage capacity by 49 %. The relative reductions are slightly smaller than for a transport distance of 100 km since the proportions of the loading and the unloading time (i.e. the components that are reduced by increasing the number of ships) relative to the roundtrip time are smaller for a transport distance of 1000 km. The gross onshore intermediate storage and ship capacities can be further reduced by approximately 27 % when offshore intermediate storage is utilised for the 3 ships scenario or by approximately 16 % when utilised for the 5 ships scenario. On the other hand, a barge with almost the same capacity as the transport ships would be required in each case.

Table 17 shows the relative increase in the most important transport chain parameters for a transport distance of 1000 km (Table 16) with respect to a transport distance of 100 km for the 20 Mt/a - Cluster scenarios (Table 15). The last three columns apply when an offshore intermediate storage is used, the first three columns when it is not.

Table 17: Relative increase in transport chain parameters for a transport distance of 1000 km with respect to the values for a transport distance of 100 km shown in Table 15. The abbreviation "off. i. st." means offshore intermediate storage.

Parameter	2 ships	3 ships	5 ships	2 ships + off. in. st.	3 ships + off. in. st.	5 ships + off. in. st.
Roundtrip time	7.1	7.1	7.2	4.1	5.2	6.0
Gross onshore intermediate storage capacity	7.4	7.7	8.1	4.3	5.6	6.7
Loading time	1.0	1.1	1.4	1.0	1.1	1.4
Gross ship capacity	7.1	7.1	7.2	4.1	5.2	6.0
Net total capacity of ships	7.1	7.1	7.2	4.1	5.2	6.0
Unloading time	7.1	7.1	7.2	0.1	0.1	0.2

When using 2 ships without an offshore intermediate storage, the gross onshore intermediate storage capacity is increased by a factor of 7.4 with respect to a transport distance of 100 km while the roundtrip time, the gross ship capacity, the net total ship capacity and the unloading time are increased by a factor of 7.1. For an increasing number of ships, the factors for the gross onshore intermediate storage capacity in Table 17 increase: To a factor of 7.7 for 3 ships and to a factor of 8.1 for 5 ships (both without offshore intermediate storage). This is due to the fact that the relative proportion of the loading time to the roundtrip time becomes increasingly short with increased transport distance (q. v. Figure 47 and Figure 48). Consequently, the time period between one ship leaving and the arrival of the other, and thus, the gross onshore intermediate storage capacity (q. v. eq. (3.11), is less dependent on the number of ships for longer transport distances. Therefore, the reduction in the gross intermediate storage capacity achieved by an increase in the number of ships is low in relative terms for longer transport distances. The factors in Table 17 are generally lower when offshore intermediate storage is used. Since the unloading time is shorter in absolute terms when offshore intermediate storage is used, the increase in transport distance leads to a smaller relative increase in roundtrip time, and ultimately, in the gross onshore intermediate storage and gross ship capacities.

The factors in Table 17 show that increasing the number of ships is more effective for shorter transport distances than for longer ones in terms of reducing gross onshore intermediate storage and ship capacities. On the other hand, the number of ships needs to be increased for larger distances once these capacities reach an unfeasible value. This might be the case for the 2 ships scenario as the required gross ship capacity of 214,000 t would be in the range of the largest LNG tankers available [4]. The required onshore intermediate storage capacity is in the same order of magnitude. Consequently, if both the gross onshore intermediate storage capacity and the gross ship capacity need to be reduced, an increase in the number of ships will often be necessary for longer distances.

Another option to reduce onshore intermediate storage and ship capacities is to use offshore intermediate storage. As illustrated in Figure 48c, using an offshore intermediate storage reduces the roundtrip time. As a result, both the onshore intermediate storage capacity and the ship capacity are reduced, but on the other hand a barge of similar capacity to the ships is required at the injection site (q. v. Table 16). Alternatively, if only the ship capacity must be reduced, a parallel transport chain of equal capacity could be established instead. With a parallel transport chain, the required ship capacity is distributed among several smaller ships in parallel operation while the onshore (and potentially) offshore intermediate storage capacities as well as the total loading, unloading and injection mass flow rates remain the same as for a single, large ship.

4.5 Energy Demand of the Baseline Scenarios

The energy demand of the individual transport chain components – liquefaction, boil-off gas reliquefaction and injection – has been calculated for the three baseline scenarios. The energy demand for liquefaction depends on the liquefaction cycle process design – i.e. 2-stage or 3-stage closed cycle process, base or optimised version. The 3-stage closed cycle process was chosen for the baseline scenarios as it is assumed that the savings in energy demand outweigh the additional investment costs of the 3-stage process. The results for the liquefaction energy demand of the three baseline scenarios are shown in Table 18. Results are calculated both for the 3-stage closed cycle base process as well as the optimised version based on the minimum specific energy demands presented in section 4.1.2. For the 1 Mt/a – Pure CO_2 scenario, pure CO_2 has been assumed. For the 2 Mt/a - Oxy96 scenario, the Oxy96 CO_2 stream and for the 20 Mt/a – Cluster scenario, the CL-MIN-CO2 CO_2 stream were used.

Table 18: Liquefaction energy demand for the three baseline scenarios presented in section 4.4.1. The 3-stage closed cycle base process and the optimised one are used.

Parameter	Unit	1 Mt/a - Pure CO_2	2 Mt/a - Oxy96	20 Mt/a - Cluster
Design mass flow rate	t/h	114	390	3236
Nominal mass flow rate per year	Mt/a	1	2.3	19.8
Spec. energy demand - base	kWh/t CO_2	12.6	12.9	12.7
Spec. energy demand - optimised	kWh/t CO_2	7.3	9.0	8.5
Electrical power required - base	MW	1.4	5.0	41.0
Electrical power required - optimised	MW	0.8	3.5	27.4
Electricity demand per year - base	GWh	12.6	29.9	250.9
Electricity demand per year - optimised	GWh	7.3	21.0	167.6
Thermal power transferred - base	MW	4.2	14.9	121.5
Thermal power transferred - optimised	MW	3.8	14.1	113.1

The electrical power required represents the power duty of the ammonia compressors for liquefaction of the design mass flow rate. It is the product of the design mass flow rate and the specific energy demand. The electricity demand per year, in contrast, depends on the nominal mass flow per year, which is the total mass per year that is liquefied (1 M/t/a, 2.3 Mt/a or 19.8 Mt/a). The difference between the design mass flow rate and the nominal mass flow rate results from the mode of operation: While a constant mass flow rate is assumed for the 1 Mt/a - Pure CO_2 scenario, variable CO_2 production characteristic of a hard-coal power plant is assumed for the 2 Mt/a - Oxy96 scenario (appendix A.2, Figure A 2) and variable CO_2 production from a cluster of different CO_2 emitters is assumed for the 20 Mt/ - Cluster scenario (appendix A.2, Figure A 3). The efficiency loss during part-load, which is primarily caused by the efficiency loss of the refrigerant compressors, is considered. The part-load characteristics of other components, in particular the efficiency loss of the expanders in the optimised process, are neglected. Screw compressors are assumed as refrigerant compressors, which are widely used in industrial refrigeration [97]. Efficient part-load operation for screw compressors can be achieved by varying the speed of the driving motor (variable speed drive). This control strategy leads to higher part-load efficiencies than the conventional slide valve control mechanism [98]. Variable speed drive control is used in the LNG industry [99] and would most likely also be used for large-scale CO_2 liquefaction. For the part-load operation of the baseline scenarios, the load-efficiency characteristic of an ammonia screw compressor from literature [100] with variable speed drive control is applied (appendix A.2, Figure A 1). The thermal power required depends on the liquefaction process and CO_2 stream. It represents the heat that is transferred from the CO_2 stream to the refrigerant rather than an additional heat demand.

Table 18 shows that the energy demand required for liquefaction depends on the composition of the CO_2 stream and the variable CO_2 production characteristic. When utilising the 3-stage closed cycle base process, the electrical power required by the 2 Mt/a - Oxy96 scenario is increased by a factor of approximately 3.5 in comparison to the 1 Mt/a – Pure CO_2 scenario, although the nominal mass flow rate per year, i.e. the total mass of CO_2 and volatiles liquefied per year, is only 2.3 times as high. This has less to do with the slightly higher specific energy demand for the Oxy96 CO_2 stream than with the higher design mass flow rate. In the case of the optimised 3-stage closed cycle process, the difference between the individual specific energy demands is greater. Thus, the difference between the electrical powers increases to a factor of 4.2. The electricity demand of the 2 Mt/a - Oxy96 scenario is found to be more 2.4 times (base process) or 2.9 times (optimised process) higher than the electricity demand of the 1 Mt/a – Pure CO_2 scenario. This can be explained by the lower efficiency of the liquefaction process for the Oxy96 CO_2 stream in comparison to pure CO_2 as well as the efficiency loss during part-load operation, which only occurs in the 2 Mt/a - Oxy96 scenario. In the 20 Mt/a - Cluster scenario, the variation in the mass flow rate over the year is lower and a higher load factor in terms of the pipeline mass flow rate is achieved. Moreover, the liquefaction process is more efficient for the CL-MIN-CO2 CO_2 stream than for the Oxy96 CO_2 stream. Consequently, the electrical power for both the base process and the optimised version is only approximately 8 times higher than in the 2 Mt/a - Oxy96 scenario, although the nominal mass flow rate per year is 8.5 times greater. The electricity demand is 8.4 times (base process) or 8.0 times (optimised process) higher than in the 2 Mt/a - Oxy96 scenario.

The thermal power transferred represents the total heat duty of the CO_2/Ammonia heat exchangers. It is generally higher for the base process than for the optimised process for all three considered scenarios. This can be explained by the lowering of the enthalpy during expansion in the optimised process. Thus, less heat has to be transferred from the CO_2 stream to the refrigeration cycle in the optimised process.

Boil-off gas calculations have been carried out for the three considered scenarios. Boil-off gas created in the onshore intermediate storage tanks is differentiated from the boil-off gas created in the storage tanks on the ship during transport. For the storage tanks on the ship, the strategies "reliquefaction", "no venting" and "release into atmosphere" are compared. The results can be found in Table 19. These values apply to one ship in the respective scenario. The boil-off gas mass flow rate and the total quantity for a roundtrip by the ship have been determined based on the calculation method presented in section 3.3.1. The heat ingress depends on the geometry of the tank, the number of tanks, the heat transfer properties of the tank material and the ambient conditions. The latter two are assumed to be the same for all scenarios. Consequently, the heat ingress per tank is the highest in the 1 Mt/a - Pure CO_2 scenario while the total heat ingress is the highest in the 20 Mt/a - Cluster scenario.

Table 19: Comparison of different handling strategies for boil-off gas created on the ship during transport for the three scenarios considered

Strategy	Parameter	Unit	1 Mt/a - Pure CO_2	2 Mt/a - Oxy96	20 Mt/a - Cluster
	Heat ingress	kW	4.96	20.35	162.62
	BOG rate	kg/s	0.015	0.008	0.039
	BOG rate per day	mass-%/d	0.124	0.032	0.011
	BOG amount per roundtrip	mass-%	0.055	0.014	0.005
Strategy 1: Reliquefaction	Coefficient of Performance	-	2.16	2.16	2.16
	Electrical power required	kW	2.29	9.42	75.29
	Ratio of CO_2 emissions generated to CO_2 reliquefied	%	1.0	7.8	13.4
Strategy 2: No venting	Temperature increase ΔT	K	0.10	0.18	0.09
	Pressure increase $\Delta p/\Delta K$ at -50 °C	bar/K	0.28	0.30	0.24
	Pressure increase	%	0.42	0.35	0.14
Strategy 3: Release into atmosphere	Fraction of CO_2 in BOG	mass-%	100.0	34.6	48.6
	CO_2 amount per roundtrip	t	0.589	0.112	0.728

The boil-off gas mass flow rate is calculated using the total heat ingress and the enthalpy of evaporation which depends on the CO_2 composition and therefore varies significantly between the scenarios. In a previous work [60], the BOG rate for a CO_2 ship with pure CO_2 and a total capacity of 20,000 m³ (4 x 5,000 m³) was found to be approximately 0.1 %/d, which is in line with other literature [24, 50, 59]. With 0.124 %/d, the results for the 1 Mt/a - Pure CO_2 scenario are of the same order of magnitude. The slightly higher BOG rate in comparison with literature can be explained by the smaller tank size used in this work and thus, the increased heat transfer surface per quantity of CO_2 transported. The results for the other two scenarios show that the BOG rate is significantly decreased in the presence of impurities, which is a consequence of the increased enthalpy of evaporation. As shown in section 4.2, the enthalpy of evaporation for pure CO_2 is 324 kJ/kg while it is 2413 kJ/kg for the Oxy96 CO_2 stream and 4182 kJ/kg for the CL-MIN-CO2 CO_2 stream.

For the calculation of the electrical power required for reliquefaction, the 3-stage closed cycle process is assumed. The electrical power required is calculated by dividing the heat ingress by the coefficient of performance. As the COP is the same in all scenarios, the electrical power only depends on the heat ingress and not on the CO_2 stream. When boil-off gas is reliquefied, the electrical power required is generated by the diesel engine, which causes additional CO_2 emissions. The ratio of these additionally generated CO_2 emissions to the quantity of CO_2 reliquefied significantly depends on the CO_2 stream composition. While the ratio is only 1.0 % for the 1 Mt/a - Pure CO_2 scenario, it is 13.4 % for the 20 Mt/a - Cluster scenario. Thus, the net quantity of CO_2 reliquefied is between 99 % (1 Mt/a - Pure CO_2 scenario) and 88 % (20 Mt/a - Cluster scenario), and is hence significantly impaired for CO_2 streams with higher impurity concentrations

As an alternative to reliquefaction, the CO_2 could also be retained inside the tanks when the pressure and temperature increase during transport can be accepted. First, the temperature increase is calculated based on the heat ingress, the CO_2 mass and the heat capacity. During isochoric heating, the CO_2 in the tank is in boiling state and both temperature and pressure are increased. In boiling state, the pressure is directly connected to temperature and the pressure increase can therefore be determined from the temperature increase. For simplification, a linear approximation of the boiling curve slope at -50 °C has been used to calculate the pressure increase. The values for the approximation of the slope $\Delta p/\Delta K$ are given in Table 19, along with the resulting relative pressure increase in percent. It can be seen that the pressure increase is small, with values below 0.5 %.

As a third strategy, the boil-off gas could be vented to keep the pressure in the tank constant. In this case, a certain amount of CO_2, depending on the BOG rate and the CO_2 fraction in the BOG stream, would be released into the atmosphere. It can be seen from Table 19 that the quantity of CO_2 emitted is significantly reduced when impurities are present. This can be explained by the higher enthalpy of evaporation of CO_2 streams with impurities. Moreover, only a fraction of the BOG stream consists of CO_2 (34.6 % of the Oxy96 BOG stream and 48.6 % of the CL-MIN-CO2 BOG stream), since the volatile components evaporate first. From the comparison of the three strategies presented, it can be concluded that boil-off gas reliquefaction on the ship is not sensible for a transport distance of 100 km since only a certain fraction of the BOG stream consists of CO_2 and the pressure increase is almost negligible when the BOG is retained inside the tanks. For a transport distance of 1000 km, the pressure increase is still limited to a single-digit percentage with values between 1.0 % (20 Mt/a - Cluster scenario) and 1.6 % (1 Mt/a – Pure CO_2 scenario). Even in this case, it is probably more economic to increase the design pressure of the tanks than to reliquefy the boil-off gas.

The same calculations have been carried out for the boil-off gas from the onshore intermediate storage tanks. The results can be found in Table 20. In contrast to the CO_2 tanks on the ship, the reliquefaction of boil-off gas from the onshore intermediate storage tanks does not require much additional investment as the liquefaction plant can be used. The electrical power required has been calculated with a COP of 2.16 (3-stage closed cycle reliquefaction process), although in practice, the BOG stream is added to the CO_2 stream from the pipeline so that the optimum operating parameters and the energy demand depend on the composition of the CO_2 stream (as discussed in section 4.1.2). This is neglected here as the impact of boil-off gas reliquefaction on the overall energy demand for liquefaction is small.

Table 20: Comparison of different handling strategies for boil-off gas created in the onshore intermediate storage tanks for the three scenarios considered

Parameter	Unit	1 Mt/a - Pure CO_2	2 Mt/a – Oxy96	20 Mt/a - Cluster
Heat ingress	kW	4.82	16.71	160.44
BOG rate	kg/s	0.015	0.007	0.038
BOG rate per day	mass-%/d	0.128	0.029	0.012
Coefficient of Performance (COP)	-	2.16	2.16	2.16
Electrical power required	kW	2.23	7.74	74.28

Table 20 shows that the results for boil-off gas from onshore intermediate storage tanks are similar to the results for the boil-off gas on the ship. The values are generally lower for the boil-off gas from the onshore intermediate storage as the capacity is lower than the ship capacity. The electrical power shown in Table 20 represents an estimation of the additional power required in the liquefaction plant. It was found to be in the range of 0.22 % to 0.27 % of the total liquefaction plant electrical power with an optimised 3-stage closed cycle process, thus it is comparably low. BOG reliquefaction for the *offshore* intermediate storage tanks is not necessary since the BOG can be used to maintain the tank pressure while emptying the tanks.

The electrical and thermal energy demands for injection are shown in Table 21 for a wellhead pressure of 120 bar and a seawater temperature of 6 °C. It can be seen that the electrical and thermal energy demands are directly proportional to the injection mass flow rate. At the same time, the specific electrical and thermal energy demands are almost the same for all scenarios. A decision that has to be made is whether the injection should be carried out directly from the ship or from a platform. In the 1 Mt/a – Pure CO_2 scenario, two ships with a comparably low injection mass flow rate are used. For this reason, direct injection from a ship might be the most economical option as no permanent offshore installation would be required. In the 20 Mt/a – Cluster scenario, the injection mass flow rate and the thermal energy demand are much higher. Despite this fact, only two ships are used, which means that a permanent platform is still probably more expensive than equipping both ships with an injection pump and seawater heat exchangers. In contrast, since offshore intermediate storage is presumed in the 2 Mt/a – Oxy96 scenario, injection would most likely be done from a barge or a platform carrying the storage tanks. Therefore, only one injection pump and one set of CO_2 heat exchangers would be required.

Table 21: Injection energy demand for the three scenarios considered. A wellhead pressure of 120 bar and a seawater temperature of 6 °C are assumed.

Parameter	Unit	1 Mt/a - Pure CO_2	2 Mt/a - Oxy96	20 Mt/a - Cluster
Injection mass flow rate	t/h	137	468	3883
Spec. electrical energy demand	kWh/t CO_2	3.8	3.3	3.5
Spec. thermal energy demand	kWh/t CO_2	26.5	27.5	26.9
Thermal energy provided by engine	%	6.1	5.1	5.6
Thermal energy provided by seawater	%	90.6	90.6	90.6
Additional energy demand	%	3.3	4.3	3.79
Electrical energy demand	MW	0.5	1.5	13.7
Total thermal energy demand	MW	3.6	12.9	104.5
Heat provided by seawater	MW	3.3	11.7	94.7
Heat provided by engine	MW	0.2	0.7	5.8
Additional energy demand	MW	0.1	0.6	4.0

Using the example of the 2 Mt/a – Oxy96 scenario, the energy demand of ship transport is compared to the total energy demand of the CCS process chain. The total energy demand for ship transport consists of liquefaction and injection (boil-off gas reliquefaction is neglected). In the case of the 2 Mt/a – Oxy96 scenario, the total specific energy demand is 12.3 kWh/t CO_2. For the Oxyfuel plant with a net efficiency of 36.6 %, this corresponds to an additional net efficiency penalty of 0.4 % pts. Since the Oxyfuel capture process itself is associated with a net efficiency penalty of 9 % pts. with regard to the conventional plant, the proportion of ship transport to the total net efficiency penalty for CO_2 capture and transport is only approximately 4 %.

111

4.6 Energy Demand in Comparison to Pipeline Transport

The transport chain studied in this work represents a combination of pipeline and ship transport. In this section, the energy demands of different transport modes are compared for two locations: For a CCS plant located on the coast, the energy demand of ship transport is compared to the energy demand of pipeline transport. For a CCS plant located inland (i.e. not on the coast), the energy demand of combined pipeline and ship transport (as studied in this work) is compared with the energy demand of pipeline transport. Offshore transport distances of 100 km and 1000 km are considered. For the inland CCS plant, an onshore pipeline with a distance of 300 km is assumed. A Post-Combustion capture plant (Post CO_2 stream) and an Oxyfuel capture plant (Oxy96 CO_2 stream) are studied for each location.

When the CO_2 capture plant is located at the coast, the output pressure of the plant must account for the pressure drop in the offshore pipeline which is either 100 km or 1000 km in length. According to literature, the optimum trade-off between pipeline diameter and the energy demand of pipe friction is usually found at a pressure drop of 15 Pa/m to 25 Pa/m [101]. Assuming a pressure drop of 20 Pa/m, the total pressure drop over the offshore pipeline is 20 bar (100 km) or 200 bar (1000 km). A wellhead pressure of 120 bar is assumed, thus the output pressure of the capture plant must be 140 bar for an offshore transport distance of 100 km. For a distance of 1000 km, the output pressure would need to be 320 bar if no booster pumps were used. Since existing CO_2 pipelines have a maximum pressure of 200 bar [23, 102], a pipeline pressure of 320 bar is considered to be too high for economic reasons. As a consequence, an output pressure of 220 bar is used at the capture plant (i.e. slightly above 200 bar to avoid the use of an additional booster pump) and a booster pump is utilised after a distance of 600 km to increase the pressure from 100 bar to 200 bar. The minimum pressure of the pipeline is assumed to be 100 bar to provide a sufficient margin to the critical pressure of 73 bar [102]. Pipeline transport is carried out in supercritical state since it is considered to be more efficient and more feasible than transport in liquid or gaseous state [4, 102, 103]: While transport in gaseous state leads to a high energy demand for compression, transport in liquid state requires refrigeration and is associated with various operational challenges since unintended evaporation of CO_2 would cause problems such as significant pressure and density changes [104].

For ship transport, the individual CO_2 stream transport pressure must be selected as the CO_2 capture plant output pressure (approximately 7 bar for the Post CO_2 stream and approximately 23.7 bar for the Oxy96 CO_2 stream). An optimised 3-stage closed cycle process with ammonia, similar to the process shown in Figure 20, is assumed for liquefaction. Since the CO_2 stream is already at ship transport pressure, expansion of the CO_2 stream is not necessary and thus, the CO_2 expanders cannot be used. The energy demand for injection can be calculated from the results in section 4.3. In addition to the energy required for CO_2 liquefaction and injection, energy for ship propulsion must be provided. The fuel consumption mainly depends on the ship capacity and ship speed. The fuel consumption of tankers typically ranges between app. 33 t fuel/h for a ship capacity of 47000 t deadweight [105] and app. 13 t fuel/h for a ship capacity of 8000 t deadweight [106] (marine diesel oil, ship speed of 15 knots). This corresponds to a specific fuel energy demand between app. 0.0127 kWh/t CO_2/km and app. 0.0340 kWh/t CO_2/km. In this work, the smaller ship capacity of 8000 t deadweight is assumed for a transport distance of 100 km and the larger ship capacity of 47000 t deadweight is assumed for the transport distance of 1000 km. It is assumed that only 75 % of the fuel energy is required for the return trip, since the ship tanks are empty when returning from the storage site to the harbour. Thus the above-mentioned specific fuel energy demand is multiplied by 1.75.

When the CO_2 capture plant is located inland, the output pressure of the plant must account for the pressure drop in the onshore pipeline and, in the case of pipeline-only transport, the offshore pipeline. The total distance for pipeline-only transport is 400 km or 1300 km, resulting in a total pressure drop of 80 bar or 260 bar, respectively. Thus, the output pressure of the CO_2 capture plant is 200 bar in the case of the 100 km offshore pipeline. For the 1000 km offshore pipeline, the output pressure of the CO_2 capture plant is assumed to be 160 bar and booster pumps are used to increase the pressure from 100 bar to 220 bar at the coast and from 100 bar to 200 bar after an offshore distance of 600 km. For combined pipeline and ship transport, a pressure of 100 bar must be reached after the onshore pipeline at the export terminal located at the coast. Thus, the output pressure of the CO_2 capture plant is also 160 bar in this case.

For the Post-Combustion capture plant, a 6-stage integrally-geared radial compressor with inter-cooling to 40 °C is assumed for pipeline transport as well as combined pipeline and ship transport. The polytropic efficiencies were taken from literature [107]. For ship transport, a 3-stage version of the same compressor has been assumed as a corresponding lower number of stages is necessary to obtain similar pressure ratios ($\Pi \approx 2$). For the Oxyfuel process, a single-stage gas processing unit (GPU) was modelled according to literature [108]. A CO_2 purity of 96 % and a CO_2 capture rate of 90 % at an air ingress of 2 % are assumed. The GPU model has not been optimised with the same effort as the liquefaction processes considered in this work. For this reason, the energy demand that was calculated is approximately 15 % higher than in literature [78]. However, the calculations are sufficient for relative comparisons between the energy demands of different transport chains since only the output pressure of the CO_2 pump changed and the energy demand for purification remains the same.

In Figure 49, the specific energy demands for pipeline transport and for ship transport are shown for a CO_2 capture plant that is located at the coast. For the Post CO_2 stream, the energy demand for compression from 2 bar to either 140 bar (pipeline transport) or approximately 7 bar (ship transport) is considered. For the Oxy96 CO_2 stream, the energy demand of the whole gas processing unit (GPU) is taken into account. Thus, the total energy demand for the Post CO_2 stream cannot be directly compared to the total energy demand for the Oxy96 CO_2 stream. The total energy demand in terms of primary energy input is generally higher for a Post-Combustion capture plant than for an Oxyfuel plant with 96 % CO_2 purity. To compare the specific fuel energy demand for ship propulsion to the electrical energy demands of the other processes, the specific fuel energy demand shown in Figure 49 was multiplied by a factor of 0.442, which is the average efficiency of fossil-fuel power plants in Germany in the year 2016 [109]. Therefore, the energy demand for ship propulsion shown in Figure 49 is a comparative value rather than the actual energy demand at the propeller shaft.

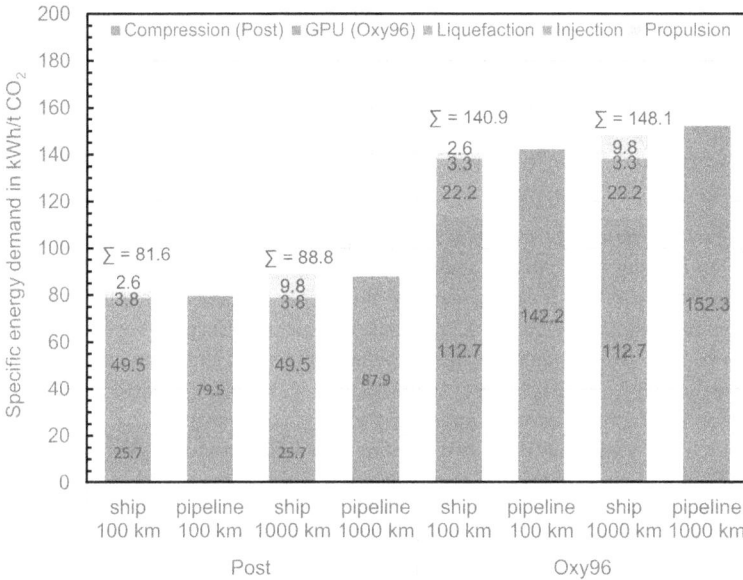

Figure 49: Specific energy demands of pipeline and ship transport for a CO_2 capture located on the coast. Offshore transport distances of 100 km and 1000 km are considered. A direct comparison between the Post and the Oxy96 CO_2 stream is not possible, since only compression has been taken into account for the Post CO_2 stream while the entire gas processing unit (GPU) has been considered for the Oxy96 CO_2 stream.

Figure 49 shows that no significant differences in regard to the specific energy demand between pipeline and ship transport can be found for a capture plant that is located at the coast: The specific energy demand for ship transport is approximately 1 % (1000 km) to 3 % (100 km) higher than for pipeline transport in the case of the Post CO_2 stream and approximately 1 % (100 km) to 3 % (1000 km) lower in the case of the Oxy96 CO_2 stream. The results demonstrate that ship transport does not generally require more energy than pipeline transport. These findings are different to those of Yoo et al. [34], who determined the energy demand of ship transport to be approximately 20 % higher than for pipeline transport. In their work, three major boundary conditions are different: For pipeline transport, a CO_2 capture plant output pressure of 110 bar instead of 140 bar (100 km) or 220 bar (1000 km) is assumed, regardless of the transport distance. Thus, the pressure drop in the pipeline is neglected. For ship transport, a 3-stage open cycle with an energy demand of approximately 61 kWh/t CO_2 is used instead of the optimised 3-stage

115

closed cycle considered in this work with approximately 50 kWh/t CO_2 (for the Post CO_2 stream). Moreover, the feed stream pressure of the liquefaction plant is assumed to be 1 bar instead of 7 bar (Post) or 23.7 bar (Oxy96). Due to these differences, Yoo et al. [34] found a lower energy demand for pipeline transport and a higher energy demand for ship transport. The differences between the results of this work and those of Yoo et al. [34] can therefore be explained by the choice of different boundary conditions.

In Figure 50, the specific energy demands of pipeline transport as well as combined pipeline and ship transport are shown for a CO_2 capture plant that is located 300 km inland from the coast. It can be seen that the energy demand of combined pipeline and ship transport is generally higher than for pipeline transport. For the Post CO_2 stream, the energy demand of combined pipeline and ship transport is approximately 12 % (1000 km) to 14 % (100 km) higher than for pipeline transport. For the Oxy96 CO_2 stream, it is approximately 7 % (1000 km) and to 8 % (100 km) higher. This means that combined pipeline and ship transport is less efficient in regard to the energy demand than pipeline transport. On the other hand, combined pipeline and ship transport might be more economic for larger distances since the investment costs of pipeline transport are generally higher than for ship transport while the additional energy demand for combined pipeline and ship transport becomes smaller in relative terms for larger distances.

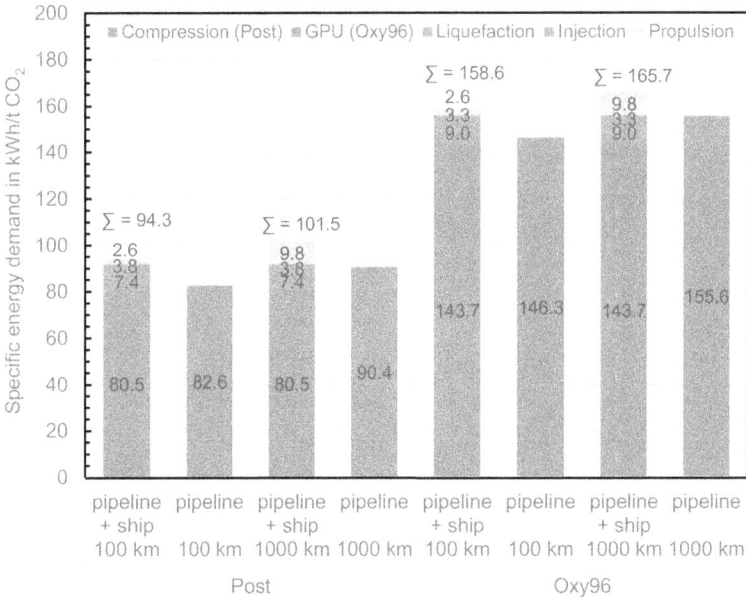

Figure 50: Specific energy demand of pipeline as well as combined pipeline and ship transport for a CO_2 capture plant that is located 300 km inland from the coast. Offshore transport distances of 100 km and 1000 km are considered.

5 Summary and Conclusions

Carbon Capture and Storage (CCS) is considered to be an effective technology to achieve significant reductions in anthropogenic CO_2 emissions. Research on CCS has historically been focused on CO_2 capture and CO_2 storage since transport between the CO_2 source and the sink is regarded as the least energy intensive and technologically challenging part of the overall CCS chain. However, efficient and reliable transport of CO_2 is critical for the successful implementation of CCS, especially in terms of public acceptance, and knowledge gaps still exist. For large-scale CO_2 transport to a potential offshore CO_2 storage location, pipeline and ship transport are both viewed as viable options. The main advantage of ship transport is its flexibility in regard to the transported CO_2 quantity and connected CO_2 sources and storage locations. Moreover, the investment costs are significantly lower compared to pipeline transport. These benefits make CO_2 ship transport an attractive option for offshore CO_2 storage, especially in the earlier stages of CCS implementation.

CO_2 is transported on ships in liquefied state, usually near the triple point at a temperature of -50 °C. Low transport pressures and high density can therby be achieved. To match the continuously liquefied CO_2 stream from the capture plant with the discontinuous ship transport, onshore intermediate storage near or at the harbour is required. When the ship arrives at the geological CO_2 storage, the pressure of the CO_2 must be raised from transport pressure to wellhead pressure and the temperature must be increased to at least 5 °C. Thus, both liquefaction and injection potentially require a certain amount of energy. During transport, a small quantity of the CO_2 is evaporated due to heat ingress from ambience into the low temperature CO_2 tanks. If this so-called boil-off gas is to be reliquefied, additional energy is required. The individual components of the transport chain – liquefaction, onshore intermediate storage, transport, boil-off gas reliquefaction (if desired) and injection are interdependent and therefore need to be considered in conjunction.

In this work, a transport chain for ship-based CO_2 transport is developed. A high-pressure CO_2 stream from a pipeline rather than a low-pressure stream from a CO_2 capture plant is assumed as the transport chain feed stream. This configuration occurs when different CO_2 emitters such as power plants, cement plants, steel plants or refineries are connected to a central pipeline for onshore transport and ship transport is desired for offshore transport to a geological CO_2 storage location.

First, the individual components - liquefaction, boil-off gas reliquefaction and injection - have been modelled in Aspen Plus® V8.6 to determine the respective specific energy demand per quantity of CO_2. For the liquefaction processes, several measurements of improvements in regard to the minimisation of the energy demand were proposed. For the injection process, different options of waste heat recovery are considered. A model of the entire transport chain was created which can be used to dimension their individual components in dependency of the transported CO_2 quantity. The transport chain model was then applied to three different scenarios in order to obtain the overall energy demand as well as the required intermediate storage and ship capacities. The impact of certain parameters such as the transport distance, ship capacity and the number of ships was studied. Contrary to other works on CO_2 ship transport, the impact of CO_2 impurities on the energy demand and the dimensioning of the individual transport chain components were analysed. Typical CO_2 compositions from Post-Combustion (Post CO_2 stream), Oxyfuel (Oxy98 and Oxy96 CO_2 stream) and Selexol™-based Pre-Combustion (Pre CO_2 stream) capture plants have been considered.

For CO_2 liquefaction, closed cycle and open cycle processes were evaluated in terms of the minimum specific energy demand per ton of CO_2. The efficiency of closed cycle processes was found to be higher than the efficiency of open cycle processes. Moreover, it was shown that open cycle processes are difficult to operate for CO_2 streams with impurities. For these reasons, only closed cycle processes were investigated in more detail. For both the 2-stage and the 3-stage closed cycle process, the minimum specific energy demand was found to be lowest for the pure CO_2 stream and highest for the CO_2 stream with the highest impurity concentration (Oxy96). However, the impact of impurities on the minimum specific energy demands was found to be small: Values between 14.6 kWh/t CO_2 (pure CO_2) and 14.9 kWh/t CO_2 (Oxy96) were found for the 2-stage closed cycle process and values between 12.6 kWh/t CO_2 (pure CO_2) and 12.9 kWh/t CO_2 (Oxy 96) for the 3-stage closed cycle process. In the next step, five measures of improvement have been investigated for the closed cycle process: Liquid and two-phase CO_2 expanders for the CO_2 stream side - a phase separator, an internal heat exchanger and an aftercooler for the refrigerant side. With all measures implemented, the minimum specific energy demand could be reduced to values between 8.6 kWh/t CO_2 (pure CO_2) and 10.8 kWh/t CO_2 (Oxy96) for the optimised 2-stage closed cycle process and values between 7.3 kWh/t CO_2 and 9.0 kWh/t CO_2 for the optimised 3-stage closed cycle process. The more significant influence of impurities on the optimised closed cycle processes can be explained by the impact of impurities on the CO_2 stream bubble pressure. Bubble pressure is the equivalence of the boiling pressure for a multi-component mixture and represents the minimum pressure that has to be maintained to ensure that the entire CO_2 stream is in liquid state. Therefore, it is the minimum pressure that is required in intermediate storage and ship tanks. For the bubble pressure, values between 6.8 bar (pure CO_2) and 23.7 bar (Oxy96) were found at a temperature of -50°C. Few literature sources were found studying the impact of bubble pressure, or impurities in general, on the liquefaction process, although bubble pressure dictates transport pressure and thus, the design pressure of all components in the transport chain.

A calculation was carried out for the mechanical design of the CO_2 ship tanks according to the DNV rules and validated against commercially available CO_2 carriers as well as literature sources on larger-scale CO_2 ship transport. As the maximum tank diameter decreases for an increased pressure, given wall thickness, and given allowable membrane stress, the corresponding tank volume also decreases for a given tank length. Thus, the investment costs per volume are significantly higher for higher tank pressures. Based on these calculations and a study of the available literature, the maximum pressure for economical ship transport was assumed to be 25 bar. Since the bubble pressure of the Pre CO_2 stream is 31.2 bar at -50 °C, ship transport of these streams was not considered further.

The injection process requires electrical (or mechanical) energy for the injection pump and thermal energy to raise the CO_2 temperature. The electrical and thermal energy demands depend on the wellhead pressure and the transport pressure of the CO_2, which in turn is determined by the composition of the CO_2 stream. The thermal energy demand increases for an increasing purity concentration while the electrical energy demand decreases. For the electrical energy demand, values between 1.9 kWh/t CO_2 (Oxy96 and a wellhead pressure of 80 bar) and 7.8 kWh/t CO_2 (pure CO_2 and a wellhead pressure of 240 bar) were found. For the thermal energy demand, values between 21.0 kWh/t CO_2 (pure CO_2 and a wellhead pressure of 240 bar) and 30.5 kWh/t CO_2 (Oxy96 and a wellhead pressure of 80 bar) were determined. Depending on the seawater temperature, most of the thermal energy demand can be provided by seawater heat. In addition to seawater heat, waste heat from the engine that drives the injection pump can be used. For seawater temperatures below 10 °C, a small amount of additional heat might be required, but is generally below 3 kWh/t CO_2.

A ship-based transport chain has been dimensioned for three exemplary scenarios with different transport capacities, feed-in characteristics and impurity concentrations: The 1 Mt/a – Pure CO_2 scenario with a constant mass flow rate, the 2 Mt/a – Oxyfuel scenario with a typical medium-load power plant feed-in characteristic and the 20 Mt/a – Cluster scenario with various different power plants and industrial CO_2 emitters. It was found that regardless of the transport capacity, two ships would most likely be the most economic option for an assumed transport distance of 100 km. For a transport distance of 1000 km, the required ship capacity in the 20 Mt/a – Cluster scenario would be so large that a higher number of ships (e.g. three to five) would be required. In addition, an offshore intermediate storage (barge) can be used to reduce both the ship capacity and the onshore intermediate storage capacity, e. g. by approximately 27 % when 3 ships are used and by approximately 16 % when 5 ships are used for a transport distance of 1000 km. The energy demand for liquefaction and injection was calculated for the three exemplary scenarios. The total specific energy demand was found to be between 11.1 kWh/t CO_2 (1 Mt/a – Pure CO_2 scenario) and 12.3 kWh/t CO_2 (2 Mt/a – Oxyfuel scenario). For the considered power plant in the 2 Mt/a – Oxy96 scenario, this corresponds to an additional net efficiency penalty of 0.4 %-pts.

The pressure increase during transport due to boil-off gas formation in the CO_2 ship tanks was found to be small. For a transport distance of 100 km, pressure increases between 0.1 % (20 Mt/a – Cluster scenario) and 0.4 % (1 Mt/a – Pure CO_2 scenario) have been determined. Thus, boil-off gas reliquefaction was found to be unnecessary for all three scenarios, most likely even for a transport distance of 1000 km and associated pressure increases of 1.0 % to 1.6 %. When accounting for the additional CO_2 emissions produced by providing energy for boil-off gas reliquefaction, the net quantity of CO_2 reliquefied would be between 99 % (1 Mt/a – Pure CO_2 scenario) and 88 % (20 Mt/a – Cluster scenario).

The transport chain studied in this work represents a combination of pipeline for onshore and ships for offshore transport. A comparison between the energy demand of pipeline and ship transport was conducted for two cases: A CCS plant located on the coast where either ship or pipeline transport is used, and a CCS plant located 300 km inland from the coast, where either pipeline transport or combined pipeline and ship transport (as studied in this work) is utilised. For a CCS plant located at the coast, the specific energy demand for ship transport was determined to be approximately 1 % (1000 km offshore transport distance) to 3 % (100 km) higher than for pipeline transport in the case of the Post CO_2 stream and approximately 1 % (100 km) to 3 % (1000 km) lower in the case of the Oxy96 CO_2 stream. These results demonstrate that ship transport does not generally require more energy than pipeline transport and thus, might be economic even for smaller transport distances. For a CCS plant located inland, the energy demand of combined pipeline and ship transport was found to be approximately 12 % (1000 km offshore transport distance) to 14 % (100 km) higher than for pipeline transport in the case of the Post CO_2 stream and approximately 7 % (1000 km) to 8 % (100 km) higher in the case the Oxy96 CO_2 stream. Despite the higher energy demand, combined pipeline and ship transport might be more economic than pipeline transport for larger distances since the investment costs of pipeline transport are generally higher than for ship transport while the additional energy demand for combined pipeline and ship transport becomes smaller in relative terms with increasing transport distance.

Literature

[1]   IPCC, "Fifth Assessment Synthesis Report-Summary for Policymakers-an Assessment of Inter-Governmental Panel on Climate Change 2014", 2014.

[2]   International Energy Agency, "Energy Technology Perspectives", 2016.

[3]   V. Scott, S. Gilfillan, N. Markusson, H. Chalmers and R.S. Haszeldine, "Last chance for carbon capture and storage", *Nature Climate Change*, vol. 3, 2013, p. 105.

[4]   R. Doctor and A. Palmer, "Transport of CO2", *Prepared by working group III of the intergovernmental panel on climate change Intergovernmental Panel on Climate Change. Cambridge, UK*, 2005.

[5]   D. Schumann, E. Duetschke and K. Pietzner, "Public perception of CO_2 offshore storage in Germany: regional differences and determinants", *Energy Procedia*, vol. 63, 2014, pp. 7096–7112.

[6]   A. Engebø and N. Ahmed, "Det Norske Veritas - Report Activity 5: CO_2 Transport", 2012.

[7]   L.E. Øi, N. Eldrup, U. Adhikari, M.H. Bentsen, J.L. Badalge and S. Yang, "Simulation and Cost Comparison of CO_2 Liquefaction", *Energy Procedia*, vol. 86, 2016, pp. 500–510.

[8]   A. Alabdulkarem, Y. Hwang and R. Radermacher, "Development of CO_2 liquefaction cycles for CO_2 sequestration", *Applied Thermal Engineering*, vol. 33, 2012, pp. 144–156.

[9]   A. Aspelund and K. Jordal, "Gas conditioning—The interface between CO_2 capture and transport", *International Journal of Greenhouse Gas Control*, vol. 1, 2007, pp. 343–354.

[10]   A. Aspelund, T. Sandvik, H. Krogstad and G. De Koeijer, "Liquefaction of captured CO_2 for ship-based transport", *Proceedings of the 7th International Conference on Greenhouse Gas Control Technologies*, 2005, pp. 2545–2549.

[11]   S. Decarre, J. Berthiaud, N. Butin and J.-L. Guillaume-Combecave, "CO_2 maritime transportation", *International Journal of Greenhouse Gas Control*, vol. 4, 2010, pp. 857–864.

[12]   U. Lee, S. Yang, Y.S. Jeong, Y. Lim, C.S. Lee and C. Han, "Carbon dioxide liquefaction process for ship transportation", *Industrial & Engineering Chemistry Research*, vol. 51, 2012, pp. 15122–15131.

[13]   A. Aspelund, T. Weydahl, T. Sandvik, H. Krogstad, L. Wongraven, R. Henningsen, J. Fivelstad, N. Oma and T. Hilden, "Offshore unloading of semipressurized CO_2 to an oilfield", *Proceedings of the 7th International Conference on Greenhouse Gas Control (GHGT-7)*, 2005, pp. 2551–2554.

[14] P. Brownsort, "Ship transport of CO_2 for Enhanced Oil Recovery – Literature Survey", *Scottish Carbon Capture & Storage*, 2015.

[15] R. A. Kajiyama Omata, "Preliminary Feasibility Study on CO_2 Carrier for Ship-based CCS", *Global CCS Institute*, 2011.

[16] B.-Y. Yoo, S.-G. Lee, K. Rhee, H.-S. Na and J.-M. Park, "New CCS system integration with CO_2 carrier and liquefaction process", *Energy Procedia*, vol. 4, 2011, pp. 2308–2314.

[17] H. You, Y. Seo, C. Huh and D. Chang, "Performance Analysis of Cold Energy Recovery from CO_2 Injection in Ship-Based Carbon Capture and Storage (CCS)", *Energies*, vol. 7, 2014, pp. 7266–7281.

[18] E. Krogh, R. Nilsen and R. Henningsen, "Liquefied CO_2 injection modelling", *Energy Procedia*, vol. 23, 2012, pp. 527–555.

[19] M. Ozaki, T. Ohsumi and R. Kajiyama, "Ship-based offshore CCS featuring CO_2 shuttle ships equipped with injection facilities", *Energy Procedia*, vol. 37, 2013, pp. 3184–3190.

[20] F. Neele, R. de Kler, M. Nienoord, P. Brownsort, J. Koornneef, S. Belfroid, L. Peters, A. van Wijhe and D. Loeve, "CO_2 transport by ship: the way forward in Europe", *Energy Procedia*, vol. 114, 2017, pp. 6824–6834.

[21] I. Al-Siyabi, "Effect of impurities on CO_2 stream properties (PhD thesis)", Heriot-Watt University, 2013.

[22] C. Eickhoff, F. Neele, M. Hammer, M. DiBiagio, C. Hofstee, M. Koenen, S. Fischer, A. Isaenko, A. Brown and T. Kovacs, "IMPACTS: Economic Trade-offs for CO_2 Impurity Specification", *Energy Procedia*, vol. 63, 2014, pp. 7379–7388.

[23] A. Oosterkamp and J. Ramsen, "State-of-the-art overview of CO_2 pipeline transport with relevance to offshore pipelines", *Polytech Report No: POL-O-2007-138-A*, 2008.

[24] P. Seevam, J. Race and M. Downie, "Infrastructure and pipeline technology for carbon dioxide transport", *Developments and Innovation in Carbon Dioxide (CO_2) Capture and Storage Technology: Carbon Dioxide (CO_2) capture, transport and industrial applications*, Elsevier, 2010.

[25] S. Walspurger and H. van Dijk, "EDGAR CO_2 purity: type and quantities of impurities related to CO_2 point source and capture technology: a Literature study", *Energy Research Centre of the Netherlands (ECN)*, ECN-E-12-054, 2012.

[26] R.T. Porter, M. Fairweather, M. Pourkashanian and R.M. Woolley, "The range and level of impurities in CO_2 streams from different carbon capture sources", *International Journal of Greenhouse Gas Control*, vol. 36, 2015, pp. 161–174.

[27] A. Kather, B. Paschke and S. Kownatzki, "CO_2-Reinheit für Abscheidung und Lagerung: COORAL; Abschlussbericht des Teilprojekts: Prozessgasdefinition, Transportnetz und Korrosion des Instituts für Energietechnik der Technischen Universität Hamburg-Harburg", *TU Hamburg-Harburg Hamburg*, 2013.

[28] A. Aspelund, "Gas purification, compression, and liquefaction processes and technology for carbon dioxide transport", *Developments and Innovation in*

Carbon Dioxide CO_2 Capture and Storage Technology: Carbon Dioxide CO_2 capture, transport and industrial applications, Elsevier, 2010.

[29]   E. De Visser, C. Hendriks, M. Barrio, M.J. Mølnvik, G. de Koeijer, S. Liljemark and Y. Le Gallo, "Dynamis CO_2 quality recommendations", International Journal of Greenhouse Gas Control, vol. 2, 2008, pp. 478–484.

[30]   W. Kuckshinrichs, P. Markewitz, J. Linssen, P. Zapp, M. Peters, B. Köhler, T.E. Müller and W. Leitner, "Weltweite Innovationen bei der Entwicklung von CCS-Technologien und Möglichkeiten der Nutzung und des Recyclings von CO2", Schriften des Forschungszentrums Jülich, Reihe Energie & Umwelt, Jülich, vol. 60, 2010.

[31]   B. Walter, "Ship transport of CO2", IEA, 2004.

[32]   S. Roussanaly, A.L. Brunsvold and E.S. Hognes, "Benchmarking of CO_2 transport technologies: Part II–Offshore pipeline and shipping to an offshore site", International Journal of Greenhouse Gas Control, vol. 28, 2014, pp. 283–299.

[33]   M. Knoope, A. Ramírez and A. Faaij, "Investing in CO_2 transport infrastructure under uncertainty: A comparison between ships and pipelines", International Journal of Greenhouse Gas Control, vol. 41, 2015, pp. 174–193.

[34]   B.-Y. Yoo, D.-K. Choi, H.-J. Kim, Y.-S. Moon, H.-S. Na and S.-G. Lee, "Development of CO_2 terminal and CO_2 carrier for future commercialized CCS market", International Journal of Greenhouse Gas Control, vol. 12, 2013, pp. 323–332.

[35]   S.K. Wang, Handbook of air conditioning and refrigeration, McGraw-Hill, 2000.

[36]   G.. Hundy, A.R. Trott and T.C. Welch, Refrigeration and Air-Conditioning, Butterworth-Heinemann, UK, 2008.

[37]   Y. Seo and D. Chang, "Optimization of ship-based CCS", OCEANS, 2012-Yeosu, IEEE, 2012, pp. 1–9.

[38]   U. Lee, Y. Lim, S. Lee, J. Jung and C. Han, "CO2 storage terminal for ship transportation", Industrial & Engineering Chemistry Research, vol. 51, 2011, pp. 389–397.

[39]   Y. Seo, H. You, S. Lee, C. Huh and D. Chang, "Evaluation of CO_2 liquefaction processes for ship-based carbon capture and storage (CCS) in terms of life cycle cost (LCC) considering availability", International Journal of Greenhouse Gas Control, vol. 35, 2015, pp. 1–12.

[40]   A. Aspelund, M. Mølnvik and G. De Koeijer, "Ship Transport of CO_2 Technical Solutions and Analysis of Costs, Energy Utilization, Exergy Efficiency and CO_2 Emissions", Chemical Engineering Research and Design, vol. 84, 2006, pp. 847–855.

[41]   M. Barrio, A. Aspelund, T. Weydahl, T. Sandvik, L. Wongraven, H. Krogstad, R. Henningsen, M. Mølnvik and S. Eide, "Ship-based transport of CO2", 7th International Conference on Greenhouse Gas Control Technologies (GHGT-7), 2005.

[42]   L. Buit, W. Mallon, P. Schulze, S. Foto and G. Stienstra, CO2-Transportinfrastruktur in Deutschland – Notwendigkeit und Rahmenbedingungen bis 2050, DNV GL, 2014.

[43]   T. Boatman, S. Jones and R. Mack, "LNG Tandem Offloading System", *AIChe Spring National Meeting*, 2003.

[44]   M. Ozaki and T. Ohsumi, "CCS from multiple sources to offshore storage site complex via ship transport", *Energy Procedia*, vol. 4, 2011, pp. 2992–2999.

[45]   Yara, "New liquid CO_2 ship for Yara, URL: http://yara.com/media/news_archive/new_liquid_co2_ship_for_yara.aspx, accessed: 2018-01-12", 2015.

[46]   H.A. Haugen, N.H. Eldrup, A.M. Fatnes and E. Leren, "Commercial capture and transport of CO_2 from production of ammonia", *Energy Procedia*, vol. 114, 2017, pp. 6133–6140.

[47]   F. Neele, H.A. Haugen and R. Skagestad, "Ship transport of CO_2-breaking the CO_2-EOR deadlock", *Energy Procedia*, vol. 63, 2014, pp. 2638–2644.

[48]   CICERO, "Carbon Capture and Storage (CCS) in a Nordic perspective", *Nordic Council of Ministers*, 2007.

[49]   L. Kujanpää, J. Rauramo and A. Arasto, "Cross-border CO_2 infrastructure options for a CCS demonstration in Finland", *Energy Procedia*, vol. 4, 2011, pp. 2425–2431.

[50]   Ø. Vesterdal, A. Hovland, A. Martinsen, T. Solvin and S. Sole, "Carbon Chain Gas Carrier", *Norwegian University of Science and Technology*, 2009.

[51]   T.N. Vermeulen, "Knowledge Sharing Report – CO_2 Liquid Logistics Shipping Concept (llsc) Overall Supply Chain Optimization", *Tebodin Netherlands*, 2011.

[52]   International Maritime Organisation, "International Code for the Construction and Equipment of Ships Carrying Liquefied Gases in Bulk", 1998.

[53]   DNV, "Rules for the classification of ships - Liquefied Gas Carriers", 2013.

[54]   DNV, "Rules For Classification Of Ships / High Speed, Light Craft And Naval Surface Craft - Newbuildings Machinery And Systems – Main Class Pressure Vessels", 2011.

[55]   DNV, "Classification notes no. 31.13: Strength analysis of independent type C tanks", 2013.

[56]   I. Senjanovi, V. Slapnicar, Z. Mravak, S. Rudan and A.-M. Ljuština, "Structure design of cargo tanks in liquefied gas carriers", *International Congress of MArine Research and Transportation-ICMRT 2005*, 2005.

[57]   VdTÜV, "AD 2000-Merkblatt W10", 2016.

[58]   M. Kraack, *LNG-Aufbereitung und Tank-Systeme*, Marine Service GmbH, 2014.

[59]   B. Chu, D. Chang and H. Chung, "Optimum liquefaction fraction for boil-off gas reliquefaction system of semi-pressurized liquid CO_2 carriers based on economic evaluation", *International Journal of Greenhouse Gas Control*, vol. 10, 2012, pp. 46–55.

[60]   B. Sengül, "Modellierung der Boil-Off-Gas-Entstehung beim CO_2-Transport auf Schiffen (Bachelor's Thesis)", TUHH, 2015.

[61]   K.-D. Gerdsmeyer and I. W.H., *On-board reliquefaction for LNG ships*, Tractebel Gas Engineering, 2005.

[62] United Nations Economic Comission for Europe, "What is Boil-off?", *LNG task force meeting in Brussels*, 2011.

[63] J.K. Jones, "Reliquefaction of boil off gas", U.S. Patent 3,857,245, 1974.

[64] S.H. Jeon and M.S. Kim, "Effects of impurities on re-liquefaction system of liquefied CO_2 transport ship for CCS", *International Journal of Greenhouse Gas Control*, vol. 43, 2015, pp. 225–232.

[65] B. Paschke, "Korrosiver Einfluss von Begleitstoffen im abgetrennten CO_2 aus Kraftwerksprozessen auf Pipeline-und Verdichterwerkstoffe", 2013.

[66] S. Brown, S. Martynov, H. Mahgerefteh, M. Fairweather, R.M. Woolley, C.J. Wareing, S.A. Falle, H. Rutters, A. Niemi, Y.C. Zhang and others, "CO_2 QUEST: Techno-economic Assessment of CO_2 Quality Effect on Its Storage and Transport", *Energy Procedia*, vol. 63, 2014, pp. 2622–2629.

[67] B. Metz, O. Davidson, H. De Coninck, M. Loos and L. Meyer, "IPCC special report on carbon dioxide capture and storage. Prepared by Working Group III of the Intergovernmental Panel on Climate Change", *IPCC, Cambridge University Press: Cambridge, United Kingdom and New York, USA*, vol. 4, 2005.

[68] M. Halseid, A. Dugstad and B. Morland, "Corrosion and bulk phase reactions in CO_2 transport pipelines with impurities: review of recent published studies", *Energy Procedia*, vol. 63, 2014, pp. 2557–2569.

[69] A. Kather and S. Kownatzki, "Assessment of the different parameters affecting the CO_2 purity from coal fired oxyfuel process", *International Journal of Greenhouse Gas Control*, vol. 5, 2011, pp. S204–S209.

[70] S.-L. Kahlke, "CO_2-Abrennung in Kraftwerks- und Industrieprozessen und Transport des CO_2 in einem gemeinsamen Pipeline-System (submitted PhD thesis)", 2018.

[71] D. Stolten and V. Scherer, "Efficient carbon capture for coal power plants", 2011.

[72] C. Han, U. Zahid, J. An, K. Kim and C. Kim, "CO_2 transport: design considerations and project outlook", *Current Opinion in Chemical Engineering*, vol. 10, 2015, pp. 42–48.

[73] Zero emissions platform, *The costs of CO_2 transport*, 2011.

[74] F. Engel and A. Kather, "Conditioning of a Pipeline CO_2 Stream for Ship Transport from Various CO_2 Sources", *Energy Procedia*, vol. 114, 2017, pp. 6741–6751.

[75] Joint Research Centre (JRC), "Reference Document on the application of Best Available Techniques to Industrial Cooling Systems", 2001.

[76] J. Boston and P. Mathias, "Phase equilibria in a third-generation process simulator", *Proceedings of the 2nd International Conference on Phase Equilibria and Fluid Properties in the Chemical Process Industries, West Berlin*, 1980, pp. 823–849.

[77] N.I. Diamantonis, G. Boulougouris, D.M. Tsangaris and I.G. Economou, "A report that identifies knowledge gaps for thermodynamic properties of CO_2 mixtures with impurities", 2013.

[78]   J. Dickmeis, "Maximierung der CO_2-Abtrennung beim kohlebefeuerte Oxyfuel-Prozess mit kryogener Luftzerlegungsanlage", 2015.

[79]   R. Eggers and S. Jeschke, "Grundlagenuntersuchungen und Modellierung der Systemdaten von aufgereinigten CO_2-reichen Abgasen aus Kraftwerken mit CCS-Technologien", 2013.

[80]   S. Lovseth, G. Sakugen, J. Stang, J. Jakobsen, O. Wilhelmsen, R. Span and R. Wegge, "CO_2 Mix Project: Experimental determination of thermo-physical properties of CO_2-rich mixtures", 2013, pp. 2888–2896.

[81]   R. Span and A. J. Jäger Gernert, "Accurate thermodynamic-property models for CO_2-rich mixtures", 2013, pp. 2914–2922.

[82]   H. Li, J.P. Jakobsen, Ø. Wilhelmsen and J. Yan, "PVTxy properties of CO_2 mixtures relevant for CO_2 capture, transport and storage: review of available experimental data and theoretical models", *Applied Energy*, vol. 88, 2011, pp. 3567–3579.

[83]   M. Kanoglu, "Cryogenic turbine efficiencies", *Exergy International Journal*, vol. 3, 2001, pp. 202–208.

[84]   J. Madison, "Comprehensive Applications for LNG Expanders", *Sixth World LNG Summit*, 2005.

[85]   H. Kimmel, "Two phase expansion", *Hydrocarbon Engineering*, 2010.

[86]   H. Kimmel, "Cryogenic LNG expanders reduce natural gas liquefaction costs", *Natural gas logistics, handling and contracts*, 2011, pp. 18–22.

[87]   H.D. Baehr and K. Stephan, *Wärme-und Stoffübertragung*, Springer, 2006.

[88]   K. Mollenhauer and H. Tschöke, *Handbuch Dieselmotoren*, Springer-Verlag, 2007.

[89]   Q. Xin, *Diesel engine system design*, Elsevier, 2011.

[90]   MAN Diesel & Turbo, "Waste Heat Recovery System for Reduction of Fuel Consumption, Emission and EEDI", 2014.

[91]   J. Wang, Z. Wang and B. Sun, "Improved equation of CO_2 Joule–Thomson coefficient", *Journal of CO_2 Utilization*, vol. 19, 2017, pp. 296–307.

[92]   H. Rachford and J. Rice, "Procedure for use of electronic digital computers in calculating flash vaporization hydrocarbon equilibrium", *Journal of Petroleum Technology*, vol. 4, 1952, pp. 19–3.

[93]   R.K. Shah, A.C. Mueller and D.P. Sekulic, "Heat Exchangers, 3. Phase Change in Heat Exchanger Design", *Ullmann's Encyclopedia of Industrial Chemistry*, American Cancer Society, 2015, pp. 1–18.

[94]   D. Webb, M. Fahrner and R. Schwaab, "The relationship between the Colburn and Silver methods of condenser design", *International journal of heat and mass transfer*, vol. 39, 1996, pp. 3147–3156.

[95]   F. Engel and A. Kather, "Improvements on the liquefaction of a pipeline CO_2 stream for ship transport", *International Journal of Greenhouse Gas Control*, vol. 72, 2018, pp. 214–221.

[96]   R. Arts, A. Chadwick, O. Eiken, S. Thibeau and S. Nooner, "Ten years' experience of monitoring CO_2 injection in the Utsira Sand at Sleipner, offshore Norway", *First break*, vol. 26, 2008.

[97] G. Lorentzen, "Ammonia: an excellent alternative", *International journal of refrigeration*, vol. 11, 1988, pp. 248–252.

[98] J.J.J. Brasz, "Comparison of Part-Load Efficiency Characteristics of Screw and Centrifugal", *International Compressor Engineering Conference*, 2006.

[99] S. Mokhatab, J.Y. Mak, J.V. Valappil and D.A. Wood, *Handbook of liquefied natural gas*, Gulf Professional Publishing, 2013.

[100] J. Cosner, "Improving rotary screw compressor performance using variable speed drives", 2009.

[101] A. M.; Golshan H. & Murray Mohitpour, *Pipeline Design & Construction: A Practical Approach*, ASME Press, 2007.

[102] S. Santos, "CO_2 Transport via Pipeline and Ship", *CCOP – EPPM Workshop on CCS*, 2012.

[103] P. Noothout, F. Wiersma, O. Hurtado, D. Macdonald, J. Kemper and K. van Alphen, "CO_2 Pipeline infrastructure–lessons learnt", *Energy Procedia*, vol. 63, 2014, pp. 2481–2492.

[104] J. Serpa, J. Morbee and E. Tzimas, *Technical and economic characteristics of a CO_2 transmission pipeline infrastructure*, Publications Office, 2011.

[105] MAN Diesel & Turbo, "Propulsion of 46,000-50,000 dwt Handymax Tanker, URL: https://marine.mandieselturbo.com/docs/librariesprovider6/technical-papers/propulsion-of-46000-50000-dwt-handymax-tanker.pdf?sfvrsn=24, accessed: 2018-06-08", 2018.

[106] MAN Diesel & Turbo, "Propulsion of 7,000-10,000 dwt Small Tanker, URL: https://marine.mandieselturbo.com/docs/librariesprovider6/technical-papers/propulsion-of-7-000-10-000-dwt-small-tanker.pdf?sfvrsn=16, accessed: 2018-06-08", 2018.

[107] U. Liebenthal, "Kennzahlen zur Quantifizierung des Einflusses einer Post-Combustion CO_2-Abtrennung auf kohlebefeuerte Dampfkraftwerke (PhD thesis)", Technische Universität Hamburg-Harburg, 2013.

[108] A. Kather, R. Eggers, C. Hermsdorf, M. Klostermann, D. Köpke and K. Mieske, "Oxyfuel-Prozess für Steinkohle mit CO_2-Abscheidung", *Abschlussbericht Verbundvorhaben ADECOS*, 2009.

[109] AG Energiebilanzen, "Ausgewählte Effizienzindikatoren zur Energiebilanz Deutschland. Daten für die Jahre von 1990 bis 2016", 2017.

[110] H. Rütters, P. Amshoff, R. Bäßler, M. Barsch, D. Bettge, N. Böttcher, F. Engel, S. Fischer, P. Jaeger, S.L. Kahlke, C. Kleinickel, A. Kratzig, Q.-H. Le, C. Lempp, J. Maßmann, F. Menezes, A. Neumann, C. Ostertag-Henning, H. Pöllmann, M. Pumpa, S. Schatzmann, S. Schmitz, S. Schütz, S. Schulz, K. Svensson, T. Weger and J. l. Wolf, "CLUSTER – Abschlusssynthese (to be published in 2019)", 2019.

Appendix

A.1 Results for the ammonia compressors

Table A 1: Results for the ammonia compressors

		Pressure ratio of compressors			Spec. el. energy demand of comp. in kWh/t CO$_2$			Comp. discharge temperature in °C		
		π_{CI}	π_{CII}	π_{CII}	w_{CI}	w_{CII}	w_{CIII}	T_{Ia}	$\dfrac{T_{IIe}}{T_{IIf}}$	T_{IIIe}
1-stage	Pure	25.1			24.0			257		
	Post	25.1			24.0			257		
	Oxy98	25.1			24.4			257		
	Oxy96	25.1			24.7			257		
	CL-MIN-CO$_2$	25.1			24.3			257		
2-stage	Pure	5.0	5.8		10.6	4.0		123	93	
	Post	5.0	5.8		10.6	4.0		123	93	
	Oxy98	4.9	5.8		10.7	4.0		123	94	
	Oxy96	4.9	5.9		10.7	4.1		122	95	
	CL-MIN-CO$_2$	4.9	5.9		10.6	4.0		123	94	
3-stage	Pure	3.1	2.9	3.6	7.4	3.5	1.7	92	58	50
	Post	3.1	2.8	3.7	7.4	3.4	1.8	92	56	52
	Oxy98	3.1	2.8	3.7	7.4	3.5	1.8	92	57	53
	Oxy96	3.0	2.9	3.8	7.4	3.6	1.9	90	58	54
	CL-MIN-CO$_2$	3.2	3.0	3.5	7.6	3.6	1.6	93	59	47

A.2 Part-load calculations

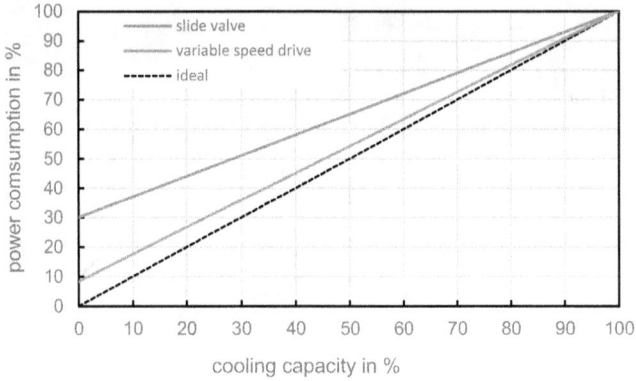

Figure A 1: Part-load characteristic of ammonia screw compressors with variable speed drive and slide valve control (data from [100])

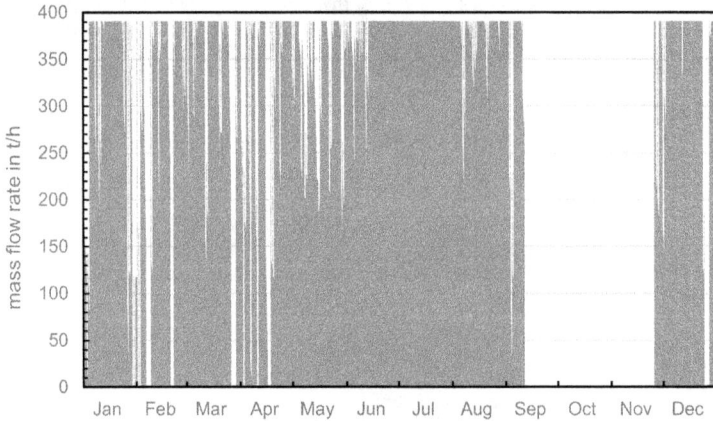

Figure A 2: Yearly CO_2 production of the 2 Mt/a – Oxy96 scenario based on the operation behaviour of a hard-coal fired power plant with Oxyfuel combustion capture technology [70]

134

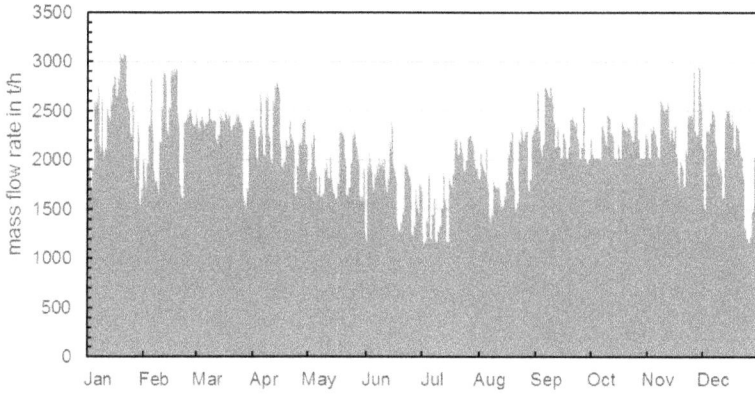

Figure A 3: Yearly CO_2 production of the 20 Mt/a – Cluster scenario based on the operations behaviour of a cluster of different CO_2 emitters [110]

Lebenslauf

Name:	Engel
Vorname:	Frithjof
Staatsangehörigkeit:	Deutsch
Geburtsdatum:	04.09.1987
Geburtsort:	Henstedt-Ulzburg

08/1994 – 06/1998	Grundschule Wöhrendamm, Großhansdorf
08/1998 – 06/2006	Emil-von-Behring-Gymnasium, Großhansdorf
08/2006 – 10/2006	Grundpraktikum im Hochbau Firma Schulz & Westphal, Ahrensburg
09/2006 – 08/2007	Studium des Bauingenieurwesens HafenCity Universität Hamburg
08/2007 – 10/2007	Grundpraktikum in der Metallbearbeitung Firma Stieger Feinwerktechnik, Ahrensburg
10/2007 – 03/2011	Studium der Energie- und Umwelttechnik Technische Universität Hamburg-Harburg Abschluss: Bachelor of Science
03/2011 – 07/2011	Praktikum bei einem Stromnetzbetreiber Firma Vattenfall Distribution, Hamburg
10/2011 – 11/2013	Studium der Energietechnik Technische Universität Hamburg-Harburg Abschluss: Master of Science
03/2013 – 09/2013	Masterarbeit an der University of Canterbury Christchurch, Neuseeland
02/2014 – 07/2019	Wissenschaftlicher Mitarbeiter Technische Universität Hamburg Institut für Energietechnik

www.ingramcontent.com/pod-product-compliance
Lightning Source LLC
Chambersburg PA
CBHW060316220326
41598CB00027B/4346